SOUND, SOCIETY AND THE GEOGRAPHY OF POPULAR MUSIC

Sound, Society and the Geography of Popular Music

Edited by

OLA JOHANSSON
University of Pittsburgh at Johnstown, USA

THOMAS L. BELL
*University of Tennessee, Knoxville, USA
and Western Kentucky University, USA*

ASHGATE

Published by
Ashgate Publishing Limited
Wey Court East
Union Road
Farnham
Surrey, GU9 7PT
England

Ashgate Publishing Company
Suite 420
101 Cherry Street
Burlington
VT 05401-4405
USA

www.ashgate.com

British Library Cataloguing in Publication Data
Sound, society and the geography of popular music.
   1. Music and geography. 2. Popular music--Social aspects.
   3. Popular music--Psychological aspects. 4. Music and
   tourism.
   I. Johansson, Ola. II. Bell, Thomas.
   306.4'842-dc22

Library of Congress Cataloging-in-Publication Data
Sound, society and the geography of popular music / [edited] by Ola Johansson
   and Thomas L. Bell.
      p. cm.
   Includes bibliographical references and index.
   ISBN 978-0-7546-7577-8 (hardback) -- ISBN 978-0-7546-9875-3 (ebook)
1. Popular music--Social aspects. 2. Music and geography. I. Johansson, Ola.
   II. Bell, Thomas L.

   ML3918.P67S68 2009
   781.6409--dc22

2009024263

ISBN 9780754675778 (hbk)
ISBN 9780754698753 (ebk)

Mixed Sources
Product group from well-managed
forests and other controlled sources
www.fsc.org Cert no. SA-COC-1565
© 1996 Forest Stewardship Council

Printed and bound in Great Britain by
MPG Books Group, UK

Contents

List of Figures

List of Tables

Notes on Contributors

Derek H. Alderman is Associate Professor of Geography at East Carolina University. He received his Ph.D. from the University of Georgia in 1998. Alderman is a cultural and historical geographer specializing in the American South. His work focuses on the changing meaning and symbolic shape of the southern landscape. While much of Alderman's published research has focused on the renaming of streets after Martin Luther King, Jr., he has a general interest in southern popular culture. For example, he has written about the geography of NASCAR, the Internet as electronic folklore, Graceland as a pilgrimage landscape, the politics of Wal-Mart, and the cultural history of kudzu.

Thomas L. Bell is a Professor Emeritus of Geography at the University of Tennessee and Adjunct Professor of Geography at Western Kentucky University. He received his Ph.D. from the University of Iowa in 1973. He has published several articles on popular culture, including "Why Seattle? An Examination of an Alternative Rock Culture Hearth" in the *Journal of Cultural Geography*. Throughout his career, Tom has also authored textbook material and research articles, primarily in the fields of economic and urban geography.

Deborah Che is Assistant Professor of Geography at Kansas State University. She received her Ph.D. from Clark University in 2000. Her research and publications have centered on rural and community development, natural resource-based tourism (i.e. agritourism, ecotourism, hunting) development and marketing, cultural/heritage tourism, and arts-based economic development strategies.

John Connell is Professor of Geography at the School of Geosciences, University of Sydney. His principal research interests are concerned with political, economic and social development in less developed countries, especially in the South Pacific region and in other small island states. Recently he has worked on the cultural geography of music and food. He has authored three books on music together with Chris Gibson.

James Craine is an Assistant Professor of Geography at California State University, Northridge. He received his Ph.D. in 2006 from San Diego State University/University of California, Santa Barbara. His current research involves the study of media geography, particularly developing new theories for the geographic engagement of visual information. He is also the editor of the *Aether*, a new online journal for media geography.

Sarah Daynes is Assistant Professor of Sociology at University of North Carolina – Greensboro. She is a social theorist, and has worked on memory, time, race, and religion. Her book *Desire for Race* was published by Cambridge University Press in 2008; in addition to various articles, her work on reggae music includes a forthcoming book, *The Politics of Hope: Time and Memory in Reggae Music*, with Manchester University Press.

John Finn is a Ph.D. candidate in the School of Geographical Sciences and Urban Planning at Arizona State University. His research interests are in cultural and musical geography, Cuba, Brazil, and social and cultural theory. He has conducted extensive fieldwork in Cuba, Brazil, and Mexico.

Chris Gibson is Associate Professor in Geography at the University of Wollongong, Australia. His research interests include the geography of creative industries, music, festivals, tourism, and transformations in place identities and economies. He is the co-author of three books on music, place and tourism and author of several academic articles and commentaries on the nexus between creative industries and regional development.

Steven Graves is Associate Professor of Geography at California State University, Northridge. He received his Ph.D. in 1999 from University of Illinois. He has researched the music industry, cultural landscapes and waste transportation issues. His research has recently focused on the spatial dynamics and the local effects of predatory lending, especially payday lending.

Edward Jackiewicz is Associate Professor of Geography at California State University, Northridge. He received his Ph.D. in 1998 from Indiana University. He has research interests in community development, housing, immigration and assimilation, local economic development, and music. He is co-editor of the book *Placing Latin America* (Rowman and Littlefield 2008).

Ola Johansson is Associate Professor in the Department of Geography at University of Pittsburgh at Johnstown. He received his Ph.D. from the University of Tennessee in 2004. His research interests include the geography of popular music, urban governance and political economy, and downtown and neighborhood redevelopment.

Sara Beth Keough is an Assistant Professor of Geography at Saginaw Valley State University in Michigan. She received her Ph.D. in Geography from the University of Tennessee. Her research focuses on globalization, media, music, identity, and migration in Canada.

Holly C. Kruse is Assistant Professor of Communication at the University of Tulsa. She is the author of *Site and Sound: Understanding Independent Music Scenes* (Peter Lang, 2003), as well as several journal articles and book chapters on communication technologies, social and economic relations, and place and space.

Robert J. Kruse II is Associate Professor of Geography at West Liberty University. He holds a Ph.D. from Kent State University (2004). Dr. Kruse's interests are in the areas of cultural geography, landscapes of popular culture, geographical aspects of disabilities, and issues of identity and space. An enthusiastic Beatles fan, his work relating to the geographies of the Beatles' career and music has appeared in his book, *A Cultural Geography of the Beatles* (Edwin Mellen Press 2005) and in articles published in *Critical Studies in Media Communication*, *Journal of Cultural Geography*, *Area*, and *Journal of Geography*. He currently teaches a course on the Geographies of Popular Culture.

Olaf Kuhlke is Associate Professor in Geography at University of Minnesota – Duluth. He holds a Ph.D. from Kent State University (2001). As a cultural geographer, he examines cultural politics and space, and the socio-spatial construction of nationalism and its expression in public spaces. He has a strong interest in the spatiality of German national identity, including German popular music. His most recent work has been in the geographies of religion. IN 2008, he published *Geographies of Freemasonry: Ritual, Lodge and City in Spatial Context* (Edwin Mellen Press).

John Lindenbaum has a Ph.D. from the University of California – Berkeley (2009). His research interests include the critical human geography of U.S. popular music, contemporary Christian music, and social change. He also sings, plays guitar and write songs for the rock bands the Lonelyhearts and Rust Belt Music.

Christopher Lukinbeal is Associate Professor of Geography at Arizona State University. He holds a Ph.D. from the joint geography program at San Diego State University/University of California, Santa Barbara. Christopher is a cultural geographer with a particular interest in cinema and media geography. His articles have appeared in *Journal of Cultural Geography, GeoJournal,* and *Journal of Geography.*

Michael W. Pesses is an Instructor of Geography at Antelope Valley College. He received a M.A. in Geography from California State University, Northridge in 2007. HIs current research interests revolve around identity and landscape, especially in the Western United States.

Kevin Romig is Assistant Professor of Geography at Texas State University – San Marcos. He holds a Ph.D. from Arizona State University (2004). His research interests are in urban sustainability, cultural productions and reactions, and popular music. He has published in *The Geographical Bulletin*, *Critical Planning*, and *City and Community*.

Leia Bell (cover image) is a poster artist based in Salt Lake City. She has been designing posters for bands professionally for the past decade (see examples on Gigposters.com or Leia Bell.com). She also operates an art gallery and studio on Broadway in Salt Lake called Signed & Numbered (signed-numbered.com).

Chapter 1

Introduction

Ola Johansson and Thomas L. Bell

Music is an integral part of the human experience. Most of us listen to music on a daily basis, either actively by choice or passively as we are exposed to music in locations where we travel, work, and relax. Some are even involved in the production of music, either as a source of recreation or as professionals. The acts of consuming or producing music are deeply meaningful to many people—they tell us a great deal about who we are, the culture in which we are embedded, and the values to which we adhere. Considering the importance that is placed on music in most cultures, it should come as no surprise that music is a legitimate topic of academic inquiry. Yet, when those of us who are engaged in research on music discuss our work, let's say in some social setting such as the proverbial cocktail party, quite a few eyebrows are raised. Sometimes that raised eyebrow indicates the opinion that music, especially popular music, has merely amusement and entertainment value and is, therefore, not worthy of further contemplation, at least of an academic nature. But more often the interest that is generated and expressed by others is genuine. That is, the endeavor of analyzing the importance and meaning of music has relevance for a lot of people beyond a small group of researchers. In any case, we are lucky to be able to think and write about music as part of our job description. As the editors of this volume we can probably speak for all the contributors, when we say that we feel strongly about the music of which we write, and hopefully that passion will be evident throughout the book as you read it.

This is a book about music and geography. The two are intimately connected and that is why, first and foremost, the volume is written from a spatial perspective. You can see in the biographical sketches of the authors that 17 of 19 who have contributed to this book are academic geographers. However, the connection between music, space, and place is interdisciplinary in nature, and the methodologies utilized by music geographers overlap to a substantial degree with approaches in sociology, cultural studies, communication, ethnomusicology, and other related disciplines. We believe, therefore, that this book will be useful to people both inside and outside geography.

This edited volume contains new research in the field of music geography. We focus especially on contemporary popular music with case studies from multiple locations around the world—the United States, the Caribbean, Canada, Australia, and Great Britain. There is an emphasis on the United States; about half of the chapters deal with aspects of American music geography. Most of the contributors are from, or living in, the United States and know its music best.

The aim of the book is to interpret the meaning of music as it pertains to spaces and places. Popular music, in this sense, is a cultural form that actively produces geographic discourses and can be used to understand broader social relations and trends, including identity, ethnicity, attachment to place, cultural economies, social activism, and politics. In no way, however, can we claim to cover comprehensively the breadth of the discipline of music geography. As an edited volume, the book is limited to the topics chosen by the individual contributors. At the same time, the authors clearly bring multiple perspectives on the relationships among music and geography; perspectives that are eclectic in terms of research methodology and underlying philosophy.

A major reason why we decided to write this book is that there are no recent edited volumes on music geography. It seemed that it was time to collect a series of new essays in a book format. The research in this book is inspired by, and indebted to, music geography and related works that have appeared during the last couple of decades. It is customary in an introduction to position a book by summarizing existing research. However, what follows is not intended to be a general overview of music geography; that has been done elsewhere in article format (Kong 1995), in encyclopedia entries (Bell forthcoming), and in introductions to special topical issues of journals (Carney 1998). Instead, the following overview introduces recent full-length books on popular music that fully or partially employ a spatial perspective. We have done this, not only to give credit to the writings that have inspired us, but also to serve as a guide for the interested reader who wants to know more about music geography.

Perhaps the most important book in contemporary music geography is John Connell and Chris Gibson's (2002) *Sound Tracks: Popular Music, Identity and Place*. As the first thorough exploration of the "new" cultural geography of popular music, it is a contemporary milestone that synthesizes existing knowledge on music and geography. It not only examines general connections between music, place, and cultural identity, but it also offers detailed perspectives on scenes, the relationship between the local and the global, and traditional geographic areas of investigation such as migration and mobility in a musical context. Given the prominent status of *Sound Tracks*, we are pleased to offer a new essay by John Connell and Chris Gibson—"Ambient Australia: Music, Meditation, and Tourist Places" in Chapter 5.

The term "identity" appears in the subtitle of *Sound Tracks*, and in fact a series of books investigates how different forms of identity are shaped by a combination of music and geographic factors. An important avenue of research is how cultural identities are formed within increasingly global cultural flows. Mark Slobin's (1993) *Subcultural Sounds* explores how musical subcultures are central to such identity formation in multiethnic Western societies. Andy Bennett's (2000) *Popular Music and Youth Culture: Music Identity and Place* contains research on the construction of young people's identity in relation to music and locality mainly from an ethnographic and sociological point of view. It investigates how music plays a role in the day-to-day lives of primarily European youth. Similarly, Martin

Stokes's (1997) *Ethnicity, Identity and Music: The Musical Construction of Place* examines how music is important in the making of identity and ethnicity. Stokes especially focuses on the role of music when regional and national identities are shaped, "postmodern" identity, the media, and evocations of place in music. Unlike *Sound, Society and the Geography of Popular Music*, Stokes's anthropologic perspective does not focus strongly on contemporary popular music. More recently Whiteley et al. have investigated the places and spaces where music is made and listened to in *Music, Space and Place: Popular Music and Cultural Identity* (2004). Discourses on identity and nationhood are explored; this time via case studies in Africa, Europe, Oceania and Caribbean. The authors provide perspectives on how music helps to communicate a shared feeling of kinship among diasporas and how rap and hip hop bring a sense of identity to minority groups, and how the music becomes a form of resistance to marginalization.

Other books have investigated distinctly geographic features of popular music. The concept of "scenes," for example, has been scrutinized. Scenes can be defined as locations where clusters of the production (artists and the music industry) and consumption (fans) of music come together to create distinct musical forms. Andy Bennett and Richard Peterson (2004) investigate music scenes and identify three archetypes—local, translocal, and virtual scenes—and provide several case studies in support of these distinctions. Some chapters in *Sound, Society and the Geography of Popular Music* further explore the changing concept of scenes as it moves away from being purely a local phenomenon to incorporating translocal and virtual elements in an increasingly interconnected world.

In Holly Kruse's (2003) *Site and Sound: Understanding Independent Music Scenes*, local "indie" scenes are investigated, especially in and around the University of Illinois at Champaign-Urbana. The author explores the cultural and economic practices of local scenes and their dialectic relationship with the musical "mainstream" and the music industry. Kruse is also a contributor to *Sound, Society and the Geography of Popular Music* (Chapter 12), where she expands on her previously published research to explore the impact of the Internet revolution on aforementioned scenes.

Several books have been dedicated to case studies on specific scenes. Sara Cohen (1991) discusses the rock culture of Liverpool and how local bands work hard to make it. Another city that is strongly defined as a center of contemporary popular music is Manchester, which has been the object of a book-length treatment in Dave Haslam's (2000) *Manchester, England: The Story of a Pop Cult City*. In the United States, Barry Shanks's (1994) *Dissonant Identities*, which focused on Austin, Texas is perhaps the most significant case study of an American local scene.

Sometimes scenes are associated with a particular genre of music. The place-based nature of several genres is investigated in this book. In a previous text, Murray Forman (2002) looked at the discourse of rap and hip hop, and how it constructs spaces, both metaphoric and concrete, such as "ghetto," "the 'hood," and the "inner city." These spaces, as socio-spatial symbols, provide especially

African-Americans with an identity both on an individual and a collective level, and they have become central to hip hop authenticity. David Knight (2006) also investigates a particular music genre—classical music—and the cultural and physical landscapes that are portrayed in the music. The research in his *Landscapes and Music: Space, Place, and Time in the World's Great Music* deviates, of course, from *Sound, Society and the Geography of Popular Music* as it emphasizes classical rather than contemporary popular music. In a US context, George Carney has, for a long time, published *The Sounds of People and Places* that covers writings in American music geography on widely different music genres. The last edition was published in 2003, and, as an anthology, it primarily contained previously published articles.

A related approach is to study the music of a particular artist. For example, the greatest pop band of all time has been subjected to a geographic analysis in Robert Kruse's (2005) *A Cultural Geography of the Beatles: Representing Landscapes as Musical Texts*. Robert Kruse is also a contributor to our book, where he builds on his previous research by examining the spatial practices of John Lennon and Yoko Ono (see Chapter 2).

Other books have connected music with traditional themes in geography. Chris Gibson and John Connell recently (2004) investigated the relationship between a central area of study in geography—tourism—and how it is related to music. Although not a geographer, Adam Krims frequently utilizes the concepts of place and space when bringing together musical and urban change in *Music and Urban Geography* (2007). Krims maps not only how cities are represented in music, but also how urbanization shape new forms and functions of music.

Lastly, we need to mention an edited text by geographers Andrew Leyshon, David Matless, and George Revill, *The Place of Music* (1998), which contributed to the study of popular music from a contemporary cultural geographic perspective. *The Place of Music* is perhaps the publication that resembles *Sound, Society and the Geography of Popular Music* most closely, although it has a more distinct British emphasis.

This overview of existing publications offers a glimpse into the wide variety of perspectives and approaches that exist in music geography. We hope this edited book on the geography of popular music, written largely by geographers, will be a welcome addition. The different chapters in *Sound, Society and the Geography of Popular Music*, building and expanding on much of this research, are also varied in character with different theoretical points of departure and different thematic content in different geographic settings. The chapters are interrelated in multiple ways, but we have chosen to present the chapters organized into six themes with two or three chapters under each theme.

The first theme ("Music, Space, and Activism") concerns the social activism of specific artists; not only how they use their music, but also various forms of media, to create "spaces of peace" (the media events of John Lennon and Yoko Ono) or "spaces of resistance" (the anti-capitalist stance of Billy Bragg).

The second theme is called "Tourism and Landscapes of Music" and emphasizes places that take on specific meanings in a sacred or quasi-religious fashion for the people who visit them. These can be pilgrimage landscapes where fans visually articulate discourses about heritage (Elvis's Graceland), or the intersection between vernacular or ethnic cultures and the global tourism industry (in Australia) where the cultural politics of representation associated with "ambient" music have the capacity to transform places both discursively and materially.

The third theme ("Mapping Musical Texts") utilizes textual or semiotic methods to analyze the memorialization of, and the meaning attached to, places as they are represented in music and lyrics. These linkages between physical landscapes, music and lyrics, and the popular perceptions of landscape are investigated in both California and the Caribbean heartland of reggae.

The fourth part of the book ("Place in Music/Music in Place") looks at the relationship between specific artists or forms of music and the places from which they originate. In Havana, Cuba, the music embedded in the cityscape can be spatialized through maps and tours. Los Angeles is seen by some commentators as the quintessential postmodern metropolis, a cultural trend that is also reflected in the music of the Red Hot Chili Peppers. And, lastly, the national identity of Canada is reflected in the music and lyrics of the Ontario-based band the Rheostatics.

The fifth theme is called "Local Music in a Connected World" and it explores the geographic unevenness in both the production and consumption of music. On the consumption side, despite globalization and cultural homogenization trends, different local musical preference patterns still exist, such as in the case of Newfoundland, Canada, where Newfoundlanders both at home and in the diaspora connect with their homeland via local music over the Internet. Scenes in various locations, such as the alternative rock scene in Champaign-Urbana, Illinois and other large cities and college towns have also been transformed through communication technologies where now translocality may be as important as traditional proximity when explaining the existence and function of local scenes.

The sixth and final focus ("The Geography of Genres") includes case studies that examine geographic perspectives on specific musical styles—hip hop, techno, and contemporary Christian pop music. In the cases of hip hop and techno, the origin of these styles is reinterpreted applying new theories and perspectives, while contemporary Christian music is approached spatially for the first time in this volume.

By organizing the chapters in this manner, we have highlighted some of the connections that exist between them; however, there are multiple connecting points as the reader will discover. In order to learn more about the individual chapters, preceding each of the six themes are introductions that we encourage the reader to turn to.

References

Bell, T.L. (forthcoming), "Geography of Music and Sound," in B. Warf (ed.).

Bennett, A. (2000), *Popular Music and Youth Culture: Music Identity and Place* (Basingstoke: Macmillan).

Bennett, A. and Peterson, R. (2004), *Music Scenes: Local, Translocal, and Virtual* (Nashville: Vanderbilt University Press).

Carney, G. (1998), "Music Geography," *Journal of Cultural Geography*, 18:2, 1-10.

Carney, G. (2003), *The Sounds of People and Places: A Geography of American Music from Country to Classical and Blues to Bop*, 4th edition (Lanham, MD: Rowman & Littlefield).

Cohen, S. (1991), *Rock Culture in Liverpool: Popular Music in the Making* (Oxford: University Press).

Connell, J. and Gibson C. (2002), *Sound Tracks: Popular Music, Identity and Place* (London: Routledge).

Forman, M. (2002), *The 'Hood Comes First: Race, Space, and Place in Rap and Hip-Hop* (Middletown, CT: Wesleyan University Press).

Gibson, C. and Connell, J. (2004), *Music and Tourism: On the Road Again* (Clevedon, UK: Multilingual Matters Limited).

Haslam, D. (2000), *Manchester, England: The Story of a Pop Cult City* (London: Fourth Estate).

Knight, D.B. (2006), *Landscapes in Music: Space, Place, and Time in the World's Great Music* (Lanham, MD: Rowman & Littlefield).

Kong, L. (1995), "Popular Music in Geographical Analysis," *Progress in Human Geography*, 19:2, 183-98.

Krims, A. (2007), *Music and Urban Geography* (London: Routledge).

Kruse, H. (2003), *Site and Sound: Understanding Independent Music Scenes* (New York: Peter Lang Publishing).

Kruse, R. (2005), *A Cultural Geography of the Beatles: Representing Landscapes as Musical Texts* (Lewiston, NY: Edwin Mellen Press).

Leyshon, A., Matless, D. and Revill, G. (1998), *The Place of Music* (New York: The Guilford Press).

Shank, B. (1994), *Dissonant Identities: The Rock'n'Roll Scene in Austin, Texas* (Middletown, CT: Wesleyan University Press).

Slobin, M. (1993), *Subcultural Sounds: Micromusics of the West* (Middletown, CT: Wesleyan University Press).

Stokes, M. (1997), *Ethnicity, Identity and Music: The Musical Construction of Place* (Oxford: Berg Publishers).

Warf, B. (ed.) (forthcoming), *Encyclopedia of Geography* (Thousand Oaks, CA: SAGE Publications).

Whiteley, S., Bennett, A. and Hawkins, S. (2004), *Music, Space and Place: Popular Music and Cultural Identity* (Aldershot, UK: Ashgate).

Part I

Music, Space, and Political Activism

In this section, we are pleased to present Robert Kruse's "Geographies of John and Yoko's 1969 Campaign for Peace: An Intersection of Celebrity, Space, Art, and Activism." Kruse has pursued research on the Beatles for a long time (Kruse 2003; Kruse 2005a) which culminated in a book about the band (Kruse 2005b). Here, he focuses on John Lennon and Yoko Ono, although not so much the music itself, but rather their public activities as celebrities.

More specifically, Robert Kruse investigates the political activism of John and Yoko and their 1969 "Campaign for Peace." A series of media-dependent events, John and Yoko's Campaign for Peace represents the most varied use of space for purposes of political activism undertaken by any prominent individuals at that time. As such, it is worthy of spatial analysis that reveals the relationship between celebrity, space, media, popular culture, art, and politics.

The chapter begins with a brief overview of some of the political geographies of 1969 followed by an identification of conceptual influences that informed the artistic work of Lennon and Ono. After a chronology of the campaign, which includes events in England, the Netherlands, Canada and other places, the author draws connections between diverse literatures of social theory, popular culture, and media to discuss the possibilities and limitations of the couple's tactics.

Kruse shows that John and Yoko understood and used the symbolism of places through their campaign events. For example, the "acorns for peace" event at the Coventry Cathedral acted as a symbolic representation of peace and reconciliation, while Amsterdam and Montreal were places selected to represent youthful liberalism. Canada especially was strategically located to be juxtaposed with the attitude of the authorities in United States, which was home to the main target audience for John and Yoko's peace message. The use of everyday spaces (e.g. the "bed-ins") emphasized the importance of individual statements and contemplation, and also via messages on outdoor advertising billboards, they conducted acts of spatial transgression by conveying private/idealistic messages in public/commercial spaces. These were novel forms of protest, which challenged traditional notions of power and resistance. The author shows how John and Yoko's media messages represented a threat to political hegemony. The blurring of boundaries between spaces of entertainment and spaces of activism by John and Yoko is still relevant today, as their strategies foreshadowed the quasi-democratic virtual media spaces of resistance.

Although the time period of study is 1969, Kruse's approach is thoroughly integrated with contemporary cultural geographic thought. For example, the role of media is increasingly explored by geographers, as evident in the growing number of sessions on the topic at professional meetings and the establishment of a new journal, *Aether*, dedicated to publish research on media geography. In fact, one of *Aether*'s co-editors, James Craine, is also a contributor to this book. Together with Edward Jackiewicz, he authored "Scales of Resistance: Billy Bragg and the Creation of Activist Spaces" which deals with, much like Kruse's article, the political activism of a particular artist. Jackiewicz and Craine argue that Bragg's music operates on multiple geographic scales, from the local level, usually addressing working-class issues in Great Britain, to the global level where Bragg offers a poignant critique of the processes associated with cultural and economic globalization, or, in the words of Jackiewicz and Craine, "multiscalar spaces of resistance."

For those not familiar with Billy Bragg, he is an English "protest singer" in the mold of Woody Guthrie and Bob Dylan, but one coming of age during the punk era. However, Bragg did not adopt punk's sometime nihilistic and exhibitionistic tendencies; rather, he realized the utility of "mixing pop and politics" and with great sincerity he has espoused an unabashedly socialist ideology in his music for more than two decades.

This chapter examines Billy Bragg's life and music in an effort to demonstrate how his music shapes cultural geographies. The authors use the lyrics and the folk-inspired music of Billy Bragg to demonstrate how he has become a voice that transcends geographic scales, moving from the local to the national to the global, sometimes in the same song. Direct activism has been central to Bragg's mission as a musician from the beginning, which is exemplified by his association with the Labour Party's "Red Wedge" campaign, but thematically, he has also evolved from traditional working-class themes to embracing multiculturalism—musically, lyrically, and ideologically. By advocating multicultural geopolitics, Bragg works to uncover the injustices of capitalist globalization.

The lyrics, sounds, performances, and actions of Billy Bragg demonstrate how music creates opportunities for activism and resistance to boundaries and borders, both social and political. Bragg's music is, therefore, a "social space" or a "space of activism," according to Jackiewicz and Craine. His musical production becomes a site of activism attacking capitalism, US hegemony, racism, and many other negative aspects of globalization. Bragg follows his convictions; creating music for him is not only a way to make a living, but also a vehicle to battle actively the injustices of globalized capitalist economies.

Jackiewicz and Craine also argue that Bragg constructs metaphoric spaces through his music. Music has tremendous affective powers on the listener, and, in the case of Bragg, he articulates socio-economic relations in society in a way to which his audience can relate. Here, the interplay between the artist and the audience is strong: it is not just about the meaning embedded within the music that needs to be teased out, but the empowerment that comes with the act of listening. Moreover, through concert experiences concrete spaces produce an inspiration and opportunity for participatory action. Bragg's activism stands in contrast to

Lennon and Ono's use of space, which focused to a great extent on their own artistic activities.

References

Kruse, R. (2003), "Imagining Strawberry Fields as a Place of Pilgrimage," *Area* 35:2, 154-62.

Kruse, R. (2005a), "The Beatles as Place Makers: Narrated Landscapes of Liverpool, England," *Journal of Cultural Geography* 22:2, 154-62.

Kruse, R. (2005b), *A Cultural Geography of the Beatles: Representing Landscapes as Musical Texts* (Lewiston, NY: Edwin Mellen Press).

Chapter 2

Geographies of John and Yoko's 1969 Campaign for Peace: An Intersection of Celebrity, Space, Art, and Activism

Robert J. Kruse II

Introduction

The Beatles continue to be of interest to scholars in the social sciences. Scholarly work on the band has applied sociological frameworks (e.g. Inglis 2000) or multi-faceted theoretical approaches blending psychology, mythology and history (e.g. McKinney 2003). As a cultural geographer, my primary interest is not the Beatles music, but rather how they have influenced the meaning of places, place representation, and tourism. My interest builds upon more general work on the spatial aspects of popular music and popular musicians that have interested cultural geographers since Larry Ford's article (1971) on the diffusion of rock music appeared more than thirty years ago. More recently, scholars have approached the spatial aspects of popular music through topics including identity (Kong 1997; Valentine 1995), place (Connell and Gibson 2003; Bowen 1997), and critical social theory (Leyshon, Matless, and Revill 1998). Specifically, the Beatles have been the focus of literature pertaining to pilgrimage (Kruse 2003) and discursive tourist landscapes (Kruse 2005a, 2005b). Common to this literature is a focus on music and musicians and their effect upon landscape interpretation. Inherent in this work, though often not discussed explicitly, are the roles that various media have played in this process.

Here, the topic is the 1969 "Campaign for Peace" conducted by then-Beatle John Lennon and his artist wife, Yoko Ono. A series of media-dependent events, John and Yoko's 1969 campaign for peace represents the most varied use of space for purposes of political activism then undertaken by any prominent individuals. As such, it is worthy of spatial analysis that reveals unique as well as generalizable relationships between celebrity, space, media, popular culture, art, and politics.

I begin with a brief overview of some of the political geographies of that year followed by an identification of conceptual influences that informed the work of Lennon and Ono in their respective fields prior to their partnership. Following a chronology of the campaign, I draw connections between diverse literatures of social theory, space, popular culture, and media to discuss the possibilities and limitations of the couple's tactics as well as the ways in which they have been recently applied by some contemporary artists. I conclude that the events, as a whole, provide an engaging case study of entangled spaces in a political context—relating to those forwarded recently by cultural geographers. Looking ahead, I propose that this is an opportune time for geographers and others in the social sciences to apply recent spatial and theoretical trends to understand the ongoing participation of celebrities and their use of media in the political arena.

Political landscapes of 1969

The political geographies of 1969 were complex, multi-scalar and inter-related. At the international scale, conventional imperial geopolitics was evident in US President Richard Nixon's policy of "Vietnamizing" the war in Southeast Asia. Elsewhere, a border dispute led to Russian and Chinese forces clashing along the Ussuri River. In arguably the most impressive example of Cold War era technologies, the geographies of humanity were extended beyond the surface of the earth to the surface of the moon. Back on earth, geographies of political protest were diffusing across borders and boundaries into different contexts.

Due to extensive media coverage, the tactics of the Civil Rights Movement in the United States were embraced by groups in other countries including the Peoples Democracy organization who, while advocating for the civil rights of Northern Ireland's Catholic minority, conducted a four-day march from Belfast to Derry—an event inspired by the march led by Rev. Martin Luther King, Jr. from Selma to Montgomery. In the United States, the boundaries between the civil rights, anti-war, black power, and feminist movements were becoming increasingly permeable. That year, the largest mass demonstration against the Vietnam War occurred on 15 November when more than 500,000 people gathered for a rally in Washington, DC. The spaces of political contest were not only international, but also local—and often activities at one scale were directly related to activities at another scale. For example, while the demonstrations at Peoples Park in Berkeley, California represented a local struggle over space, they were also representative of liberal youth opposition to hegemonic geopolitics.

In 1969, the Beatles remained among the largest influences on popular culture, music and style. The preceding year had been a turning point for John Lennon, both in terms of his personal life and the direction his work would take. Due, in part, to the 1968 student protests in Paris, Lennon was ambivalently drawn into the politics of the New Left. In the same year, he met and began a personal and artistic partnership with avant guard artist Yoko Ono. Together they began to focus their work on the promotion of "peace." Beginning with a 1968 exhibit called "Acorns for Peace," the couple conceived of a multi-faceted campaign that included events, performances, media manipulation, and advertising. In the process, due to their fame and the perceived political influence of popular culture on youth politics, they became involved in some of the prominent issues on a larger, contested global political landscape.

The politicization of John Lennon and Yoko Ono

By 1968, the Beatles' music was increasingly scrutinized in terms of political statements. That year the Beatles released Lennon's song "Revolution" as a single. Lennon's lyric, while clearly left-leaning, also expressed ambivalence towards leftist political tactics. Consequently, he was criticized by some groups for not taking a clear political stand, something many felt he had an obligation to do because of his influence on youth. For example, Tariq Ali, editor of the leftist newspaper, *Black Dwarf*, and organizer of marches on the American embassy in London, expressed frustration at Lennon's lack of political involvement. Lennon responded to Ali's editorial by writing, "I'm not only up against the establishment, but you, too" (quoted in Wiener 1984, 82). Later in the year, Lennon responded in a more conciliatory tone to an article in the newspaper that was critical of his politics. He wrote that the newspaper and he were making different contributions to the same cause (Dworkin 1997). On one hand the ambivalence and ambiguity of Lennon's political statements of this era suggest that he was unclear—still in the process of determining his position and role, if any, with regard to contemporary leftist politics. On the other hand, it appears that he wanted the respect of left-leaning activists and to maintain his image as the intellectual Beatle. Whatever his reasons—criticism from the left, excitement over his collaborations and romance with Yoko Ono or his own vanity, Lennon's work for the next two years expanded beyond the boundaries of popular music into the conceptual and philosophical contexts of avant guard art—in the name of "peace."

Without a doubt, Lennon's ability to attract the attention of the media was a key element of the campaign for peace. However, the spatial tactics of the campaign were influenced more by Ono's approach to art than Lennon's position as a rock music celebrity. As a college student in Japan, she had been a philosophy major at Bakushuin University. Although she withdrew after two semesters, she encountered the dominant intellectual trends of postwar Japan, especially Marxism and existentialism. After moving to New York City in 1955, Ono became immersed in the city's avant guard art scene which was drawing upon diverse cultures and landscapes in its works. At that time, the downtown art scene was experimenting with combinations of East Asian aesthetics, poetry and Euro-American modernism. Munroe (2000, 16) describes the work of artists with whom Ono interacted as:

> reacting against the heroics of Abstract Expressionism and the commercialism of high modernism. The new movements – neo-Dada, assemblage, Happenings and Fluxus – championed anarcho cultural sensibilities drawn from Dada, Western Phenomenology and existentialism, and notions of minimalism, indeterminacy, and everyday realism extracted from Buddhist thought.

Yoko became involved with a group of experimental artists known as Fluxus, whose members challenged societal norms by expressing "everyday life and its natural, often humorous relation to art" (Munroe 2000, 17). Experimenting with performance, film, painting and installations, Ono created a body of work that involved the use of recycled objects "fabricated from everyday, non-art materials … whose form [was] radically reductive" (Munroe 2000, 31).

The worlds from which Ono and Lennon emerged both worked to recontextualize the spaces of everyday life. The re-setting of everyday life was especially evident on the Beatles' album, *Sgt. Pepper's Lonely Hearts Club Band*. For example, the songs "Good Morning, Good Morning" and "A Day in the Life" both defamiliarized mundane events and highlighted peculiarities of the ordinary. My use of the term "everyday life" refers to the condition of modern living described by critical theorists including Henri Lefebvre and applied to popular culture by Lawrence Grossberg. In cultural terms, Grossberg (1997) suggests that rock music offers its audience the possibility, through affect and expressions of desire, of escaping the inherent boredom of modernity, while remaining a function of everyday life. Similarly, the traditions of Ono's work deconstructed and recontextualized features of everyday life. Hence, although Lennon's work was commercial and Ono's was not, the couple shared an interest in representations of everyday life that would be an important influence on their choice of spatial tactics for the peace campaign. Part of their strategy was to play with the dominant cultural discourses that referred to the spaces of everyday life.

Chronology of the peace campaign (1968-1969) *Acorns for Peace*, national sculpture exhibition, Coventry Cathedral, England

Foreshadowing their formally announced peace campaign, Lennon and Ono participated in the National Sculpture Exhibition on the grounds of Coventry Cathedral in England on 14 June 1968. Although an awkward and poorly-coordinated event, it previewed spatial tactics that Lennon and Ono would develop throughout 1969 during their campaign for peace. I maintain that the event indicates, to some degree, the couple's understandings of the symbolism of places, and the symbolic potency of juxtaposing "nature" with modernity. It also employed features of performance art such as the use of everyday (non-performance) spaces, and emphasized the importance of individual statements and contemplation. Moreover, media coverage was integral to the concept of the exhibit.

Lennon and Ono chose to participate, in part, because of the Cathedral's symbolism as a place of reconciliation and peace. Rebuilt after Coventry was bombed in 1940, the Cathedral, under the direction of Provost Dick Howard, has provided spiritual and practical support to war-torn places throughout the world (Coventry Cathedral 2003). The couple's exhibit involved their planting two acorns, each in a plastic cup, and erecting a circular bench around the seedlings at which visitors could contemplate world peace. After arriving at the Cathedral grounds, John and Yoko quickly dug a hole and planted one acorn facing East and the other facing West. After Lennon made a brief statement to attending reporters on the symbolism of their installation for peace between the East and the West, the couple departed in their white limousine. Although their planting of acorns was a seemingly simple and innocent statement, the site of "Acorns for Peace" quickly became a contested space. After Lennon and Ono departed, their exhibit was moved from its original location in front of the Chapel of Unity to the cathedral gardens. Reportedly, Canon Stephen Verney ordered the piece moved because he objected to the use of consecrated ground for the couple's work because of their extra-martial relationship (Chambers 2005). Shortly thereafter and adding further complication to the exhibit, the acorns were removed from the cathedral grounds by a thief.

Bed-in for peace (1), Amsterdam, the Netherlands

John Lennon and Yoko Ono were married in Gibraltar on 20 March 1969. Aware that their marriage would be covered by the world's press, they used the occasion to formally launch a "Campaign for Peace." Yoko Ono describes the couples' spatial tactics:

> We worked for three months thinking out the most functional approach to boosting peace before we got married … [We decided to spend] our honeymoon talking to the press in bed in Amsterdam. For us it was the only way. We can't go out in Trafalgar Square because it would create a riot. We can't lead a parade or a march because of all the autograph hunters. We had to find our own way of doing it and for now, bed-ins seems to be the most logical way. We think the bed-in can be effective (quoted in Fawcett 1976, 45).

The couple invited the world's press to their honeymoon suite in room 902 of the Hilton Hotel in Amsterdam to promote "peace." Their choice of Amsterdam was not arbitrary. As wealthy celebrities, they had considerable, but not unlimited, mobility and could have chosen almost any location. Amsterdam was important because of what it did and did not symbolize. For example, while the couple could have chosen London, it is evident that Lennon wanted to downplay his nationality. Lennon explains his ambivalence toward Britain:

> We'd be happier if we could say, "We, the British, aren't doing anything [to promote violence]; we're a peaceful nation." The drag about being British is that we're still imperialist even though we haven't got the power (quoted in Guiliano and Guiliano 1998, 46).

Moreover, Amsterdam's reputation as a place of youthful liberalism was consonant with the message the couple was trying to convey. Hence, the media attention provided Lennon and Ono with the opportunity to represent the city of Amsterdam in a positive light that reinforced the message of their campaign. At a press conference, the couple explains their choice of Amsterdam as the most appropriate place for their purposes:

John: It's a beautiful place.

Yoko: A romantic place.

John: There's a few centers in the world and Amsterdam is one of them for youth.

Yoko: Many vitally alive youth with high ideals ... for the world (quoted in Giuliano and Devi 2000, 120).

Without providing a detailed strategy for world peace or policies, John and Yoko met with members of the media and discussed, in general terms often laced with humor and irony, the need for a non-violent and peaceful world. Their stated intent was to make peace and non-violence topical by capitalizing on the media's fascination with them as celebrities. The photographs published worldwide at the time show the couple in their pajamas seated on a large bed. At the very least, through the press and television coverage, the event drew attention to the couple and left the impression that Lennon and Ono were exposing their privacy to talk about peace.

Bed-in for peace (2), **Montreal, Canada**

A few months later, after briefly considering hosting a second bed-in in the Bahamas, Lennon and Ono chose the city of Montreal. Lennon, writing nearly a decade after the event, explained the circumstances:

> We tried to repeat our great success in America, by taking the show to Broadway (the Plaza, actually). But the U.S. government decided we were too dangerous to have around in a hotel bed talking about peace. So we took the act to Montreal and broadcast (by radio and TV) across the border (Lennon 1986, 22).

The dynamics of the second bed-in were quite different from those of the first one. Especially significant was the issue of exclusion. Due to a conviction for possession of marijuana in England, Lennon's visa application to the United States was denied. Lennon publicly contextualized the denial in terms of the Cold War and stated, "It's easier to get into Russia than the U.S" (quoted. in Wiener 1984, 91). The couple complimented Canada publicly for allowing them to enter the country. In addition to capitalizing on the liberal atmosphere of Canada, the couple understood the city's strategic location in relation to the media industry of the United States, their target audience. Their Montreal bed-in was the first of three visits to Canada the couple made that year. As they had done in the Netherlands, John and Yoko used the Montreal bed-in as an opportunity to represent their host country to the world in a flattering light, and in contrast to the United States. Lennon told the world that Canada was the first country to help with their peace campaign and that he and Yoko were astonished at how well Canadian reporters had treated them.

When talking with the press during the bed-in, Lennon repeatedly took the opportunity to throw jabs at the US government. Asked why the couple was not talking directly with policy makers and power brokers, he responded, "Shit, talk? Talk about what? It doesn't happen like that. In the US, the government is too busy talking about how to keep me out. If I'm a joke, as they say, and not important, why don't they just let me in?" (Fawcett 1976, 53). However, while the couple criticized the US government, they openly embraced the powerful form of mass-marketing unique to American capitalism. Lennon explained:

> Henry Ford knew how to sell cars by advertising. I'm selling peace, and Yoko and I are just one big advertising campaign. It may make people laugh, but it may make them think, too. Really, we're Mr. and Mrs. Peace (quoted in Fawcett 1976, 54).

Responses by the American media to the bed-in and the couple's peace campaign were mixed. Newspaper headlines included "This must rank as the most self-indulgent demonstration of all time" and "Beatle Lennon and his charmer Yoko have now established themselves as the outstanding nut cases of the world" (Hopkins 1986, 100). In contrast, *Rolling Stone* magazine suggested that "A five-hour talk between John Lennon and Richard Nixon would be more significant than any Geneva Summit Conference between the U.S.A and Russia" (Norman 1981, 388).

During their week in Montreal, John and Yoko delivered live radio messages to participants of an ongoing protest at People's Park in Berkeley, California. In contrast to the light-hearted symbolism of the Amsterdam bed-in, the couple used the situation at People's Park to claim that they were, in fact, having a positive, observable effect on the political landscape. Lennon explained:

> The best thing we've done so far is talk with people at Berkeley and we think we had some influence in holding back any violence that's going to come out of that (CBC 2000).

Another significant difference between the Amsterdam and the Montreal bed-ins involved the recording of Lennon's song, "Give Peace a Chance". On 1 June Lennon and Ono, accompanied by a group of celebrities including Timothy Leary, radio personality Murray the K., Tommy Smothers, Canadian peace activist Rabbi Feinstein, and members of the Radha Krishna Temple, recorded the song which was released shortly thereafter. Possibly the most influential component of the bed-in, and maybe of the entire campaign, the song was sung later in the year by hundreds of thousands of protestors at the Vietnam Moratorium Day march in Washington DC on 1 October 1969.

Acorns for peace (2), (global scale)

The couple expanded their "Acorns for Peace" from the localized exhibition at Coventry in 1968 to one global in scope in 1969. Despite the change in scale, the couple still intended to equate peace with the symbolic act of tree planting, this time with the participation of world leaders. Hence, in the months between the Amsterdam and Montreal bed-ins, the couple began sending "Acorns for Peace" to many heads of state of the world's nations. Emphasizing the receptivity of leaders of non-super power nations, Lennon recalled in 1970:

> We sent one to Harold Wilson. I don't think we got a reply from Harold … I believe Golda Meir said, "I don't know who they are, but if it's for peace, we're for it" or something life that. [In] Scandinavia somebody or other planted it. I think Laile Selassie planted his … There were quite a few people that understood the idea (quoted in Wenner 1971, 178).

Live Peace in Toronto, **Varsity Stadium, University of Toronto**

Lennon and Ono also utilized the format of the stadium rock concert, a venue pioneered by the Beatles when they were the first major group to perform in sports complexes and stadiums in North America. The purpose at the time had been to increase the profitability of their geographically expansive tours of the United States and Canada. During their second trip to Canada, John and Yoko performed with a transient group of musicians they called "The Plastic Ono Band" at a rock and roll revival show held at Varsity Stadium at the University of Toronto. While the show included stars such as Little Richard, Lennon and Ono spun their participation as another manifestation of their peace campaign. A recording of the concert was subsequently released under the title, "Live Peace in Toronto." As they had with "Acorns for Peace," they made efforts to globalize live performances for peace by planning the largest peace/music festival that would tour globally. The unrealized festival, which was intended to inaugurate 1970 as "year one for peace," would have been unprecedented in scale and mobility.

In interviews and press conferences during their third visit to Canada, the couple continued to represent the country in progressive, liberal terms to the world's media. Quick to point out that the plans for the huge, mobile peace festival came from Canadian promoters; Lennon contrasted the attitudes in Canada with those in the United States:

> The idea came to us from Canada. It's a young country. It's right next door to the other place and you haven't been converted (CBC 2000).

Allan Rock, a former member of the Canadian parliament who was involved with the couple's Canadian events, put the importance of Canada to the peace campaign in historical terms:

> We were Canada. We were, as we'd always been, that middle ground between the old world and the excess of America. It made it an attractive place [for Lennon] (CBC 2000).

While Lennon and Ono were in Canada for the third time in 1969, Pierre Trudeau agreed to meet with the couple to discuss their plans for the peace festival. Probably with mixed motives that included the political benefits of being photographed with the famous couple, Trudeau met with Lennon and Ono and offered the official support of the Canadian government for the festival. Although Trudeau, Lennon, and Ono were from very different cultural realms, Trudeau had been represented in the media as "hip" and in tune with young people during a recent trip to London. Also, consistent with Lennon and Ono's emphasis on dreaming peace, Trudeau had spoken of dreaming as an instrument of social change. For example, in 1968 Trudeau was asked during an interview if he had a "vision" for Canada.

> I dream. I dream all the time. I've always dreamt of a society where each person would be able to fulfill himself to the extent of his capabilities as a human being, a society where inhibitions to equality would be eradicated. This means providing individual freedoms, and equality of opportunity, health, and education, and I conceive of politics as a series of decisions to create this society (quoted in Iglauer 2001, 56).

John and Yoko's meeting with Trudeau briefly moved their peace campaign from the margins of establishment politics to its center. Notably, Lennon was photographed wearing a conservative jacket and tie to the meeting with the prime minister. Like Trudeau, Lennon and Ono understood the potential benefits of such an alliance. Recalling the meeting, writer Ritchie York notes that, while the couple publicly challenged the dominant political discourses of the establishment, they also sought to be taken seriously by it:

> [John] had to get through to a prominent politician to lend credibility to his movement and, of course, we were hoping to try and wrangle an endorsement. Trudeau was willing to meet them as long as there was no advance publicity, so [they] did it and then announced it afterward (quoted in Giuliano and Devi 1999, 279).

After the cordial meeting, Lennon told the press:

> If there were more leaders like Trudeau, the world would have peace … You don't know how lucky you are in Canada (quoted in Fawcett 1976, 67).

Outdoor advertising in global cities, *War is Over If You Want It* (billboards in Paris, Rome, Berlin, Athens, Tokyo, New York, Los Angeles, Toronto, Montreal, Port-of-Spain, Trinidad)

Utilizing the conventions of outdoor mass advertising, Lennon and Ono bought billboard space in major cities. Translated into native languages, each billboard showed the same phrase printed in black letters on a white background, "War is Over – If You Want It. Happy Xmas from John and Yoko." The same phrase was reproduced as posters that were to be distributed by volunteers in suburban areas. Notably, the campaign was centered in European and North American cities. Canadian lawyer Mark Starowicz remembers the billboards:

> I cannot emphasize enough the scale of the billboards. It was as big as if you were launching a new Nike product or something (CBC 2000).

Linking their commercial-styled billboard campaign to the private spaces of everyday life, Lennon suggested that the idea could be realized at different scales and in different spaces:

> People can do more posters. For instance, locally, you know. Everybody says, "It's all right for you two, you have all the facilities." If some housewife is against war – [she could] just put it in the window – just to let her neighbors know or her husband know. You've got to start on the home ground, you know. Convert your parents (CBC 2000).

Consistent with their strategy of spatial transgression, John and Yoko's use of a conventional form of outdoor, visual advertising, effectively placed their peace message in the ordinarily commercialized spaces of millions of city dwellers.

Yoko has continued to use outdoor urban billboards occasionally to promote peace and non-violence. For example, in December 2000, the 31st anniversary of the billboard campaign and the month of the 11th anniversary of Lennon's murder, she commissioned a billboard in Times Square, New York City that read: "Over 676,000 people have been killed by guns in the U.S.A. since John Lennon was shot and killed on December 8, 1980. Commissioned by Yoko Ono December 2000."

The campaign: A spatial overview

Novel, insightful, artistic, silly, or pretentious, the events of the peace campaign were and remain unconventional resistant political statements. Lennon and Ono drew upon their fame, art, feminism, telecommunications, and everyday life with a lightness of touch and humor that was not evident in the other examples of resistance of that era. As such, one could view the events and performances as a series of peace poems in various media—a style that does not lend itself easily to the dominant assumptions of conventional spatial analysis. Routledge (1997, 68) describes the challenge of theorizing such forms of resistance: "Attempts to theorize resistance have been fraught with an intellectual taming that transforms the poetry and intensity of resistance into the dull prose of rationality." By proposing that the cause of peace would be furthered if "everyone went back to bed" and "dreamed" peace, Lennon and Ono were conceptualizing an interplay of space, power, and political resistance evident in recent geographical literature. Such tactics challenge structuralist understandings of power and resistance (Knopp 1997; Thrift 1997). As Pile (1997, 16) writes:

> Resistance ... cannot simply address itself to changing external physical space, but must also engage the colonized spaces of people's inner worlds. Indeed, it could be argued that the production of "inner spaces" marks out the real break point of political struggle ... maybe.

Relatedly, the spatial tactics of the peace campaign included the "classes of life [such as] dreams, daydreams, and fantasies and practices of 'creative proliferation' [that] are increasing studied in spatial terms by geographers" (Thrift 2003, 273).

The campaign was symbolic, mobile and without a spatial focus. Although the US government denied Lennon and Ono physical access to the United States, the couple's deliberate manipulation of the media rendered the attempt unsuccessful in political terms. Moreover, the campaign operated outside of the familiar and predictable spaces of protest. To those paying attention, it unfolded throughout the year as a series of surprises—of unpredictable choices of symbolic and strategic spaces. In addition, as it relied on media coverage, it was virtually impossible to constrain spatially. Such tactics can shift the power/resistance dynamic and can represent a threat to hegemonic political authority. As Pile (1997, 3) explains:

> One of authority's most insidious effects may well be to confine definitions of resistance to only those that appear to oppose it directly, in the open, where it can be made and seen to fail.

Geographical particularities of the events

Whether deliberate or not, the couple's first event, "Acorns for Peace," relates to long-held cultural associations of trees and garden landscapes with the concept of idealized peace and paradise. For example, the links between gardens, planting and peace are well-documented in English literature. Sir Francis Bacon and English poets including Chaucer, Shakespeare, Herbert, and Milton wrote of the peaceful healing and spiritual lessons they found in their gardens (Brother 1956).

The tradition of tree planting as a symbolic act for peace is evident in contemporary political statements as well. For example, Break the Silence, a network of American Jews who support a peaceful settlement to the Israeli-

Palestinian conflict, conducted an "Olive Trees for Peace Campaign" which they announced in a full-page advertisement in *The New York Times* on the 8 April 2001 (*Washington Report on Middle East Affairs*, May 2001). Relatedly, in 2005 the Nobel Committee honored Kenyan activist, Wangri Maathai, for her advocacy of peace and the campaign she led that included women volunteers planting more than 30 million trees in Kenya (Ahmad 2005).

Although John and Yoko's use of the act of planting on the landscape as symbolic of peace has many historical precedents, it was not a common tactic of political resistance in 1969. Notably, 37 years after the couple's exhibit at Coventry, Yoko Ono continued to employ the symbolic planting of trees for peace. Returning to Coventry in 2005 in recognition of her first peace collaboration with John Lennon, Ono gave an oak sapling to each school in the city. Finally achieving the anonymous aesthetic of Fluxus, Ono was unknown to most of the children—most of whom viewed her simply as a 72-year-old woman who donated the trees to their schools (Crutchlow 2005).

More widely covered by the press and of longer duration were the bed-ins. Lennon and Ono were well aware that the media would be attracted to a room where two celebrities were to be in bed. In an era that challenged conventional notions of sexuality, the sight of Anglo John and Asian Yoko in bed together would be tantalizing. Moreover, the image of the couple, when contextualized in terms of the Vietnam War and homogenized media representations of "Orientals," was especially provocative. However, looking past the association with bedrooms and sex, a more serious symbolic engagement with discourses of bedrooms is evident. First, the bedroom was an appropriate place to convey their primary message—that peace was available through the imagining and dreaming of individuals located in places traditionally viewed as private. Hence, the association of bedrooms with dreaming and playfulness worked well to further the couple's discourse on the prospects for peace. Yoko describes the couple's promotion of dreaming as instrumental for positive social change in a hierarchy of places:

> As John and I used to say, "A dream becomes a reality when two people dream together. Little by little things can change. First be good parents. Improve your city. Create a better country, then, finally a better world (quoted in Giuliano and Devi 1999, 79).

Elliott further elaborates on the symbolic choice of the bedroom:

> Throughout the bed-in, Lennon and Ono remained on their bed: the bed itself became a site for dreaming politics anew and also for coping with the overwhelming anxieties provoked by a militarized world (Elliott 1999, 126).

This conceptualization of dreaming and dream spaces can be linked to representations of space offered earlier in the twentieth century by surrealist artists who, in some cases, viewed such representations as political acts. Notably, Yoko

Ono's background as an artist included a familiarity with surrealist aesthetics which she evidently shared with John Lennon. Marcuse (1955, 149) describes the fundamental question posed by surrealist artists: "'Cannot the dream also be applied to the solution of the fundamental problems of life?' They went beyond psychoanalysis in demanding that the dream be made into reality without compromise." Similarly, Benjamin (1986, 185) describes how the representations of space evident in surrealism moved from being a "highly contemplative attitude" to a "revolutionary opposition" to the dominant discourses of modernity following World War I.

In some respects, the photographs and film taken by the media could have been interpreted as depicting scenes from the couple's everyday life. Though the bed-ins were performances, Lennon and Ono depended upon the media to represent them as domestic. Such an approach is consistent with Routledge's (1997, 361) observations:

> Resistances are assembled out of the materials and practices of everyday life and imply some form of contestation, some juxtaposition of forces involving all or any of the following: symbolic meanings, communicative processes, political discourses, religious idioms, cultural practices, social networks, physical settings, and bodily practices and envisioned desire and hope.

The couple's performances linking the process of world peace to the domestic spaces of everyday life have roots in feminist theory that challenges the public/private spatial dichotomy that Domosh (1999) and others have described. Moreover, when one adds the influence of media and telecommunications to the public/private spatial dichotomy, the impossibility of a distinct boundary between the two realms becomes apparent and the possibilities of political involvement in everyday life increase. As Meyrowtiz (1985, 223) writes, "Television may be the common denominator linking the 'politicizing of the personal' and the 'personalizing of the political.'"

Notably, John and Yoko were invoking *particular* western cultural discourses pertaining to bedrooms, especially that of the contemporary bedroom as a feminized space, a response to the masculinized corporate spaces in which so many people in the industrialized countries conduct so much of their everyday lives (Mirer 2000). However, there are indications that the bedroom as place has changed discursively since 1969. For example contemporary bedrooms are increasingly multi-purpose spaces that include en-suite bathrooms, computer areas and elaborate luxury (*The Journal* 2005). Moreover, Mirer (2000) adds that contemporary bedroom designs may represent repressed class anxieties which have seeped from the public realm into private spaces. By blurring these boundaries, bedroom designs may reflect the diminished privacy and increased "productivity" of home spaces characteristic of modern life.

In terms of dominant geopolitical discourses of the era, Lennon and Ono could play with conventional uses of places for the purposes of their peace campaign, but they could never escape the hegemony of the Cold War. Though they played the superpowers against each other rhetorically, in doing so, they simultaneously reinforced the hegemonic geopolitical environment. Ironically, while showing some sensitivities to British colonialism (e.g. Lennon's explanation of their choice of Amsterdam), the couple seemed oblivious to possible symbolic contradictions, especially during their bed-ins. Intending to communicate domesticity, privacy and bed-room attire, the pajamas they wore could also have been interpreted as symbols of British colonialism (pajamas were popularized as nightwear in the West by the British colonization of India).

Just as television had played a major role in the cultural impact of rock and roll music, its presentational strengths highlighted the affective, rather than instrumental sort of power the couple was proposing. Moreover, John and Yoko's peace campaign was able to join the many political messages circulating through the media—not because of the gradually accrued political capital or credibility by the couple, but because of the way television integrates various types of information (Meyrowitz 1985)—the serious and the silly, the private and the public, popular and high culture, commercialism and politics. Hence, Lennon and Ono were able to make it appear that peace was the topic of the moment—an example of television's ability, through images, to make political protests appear as social realities (Meyrowitz 1985).

Through what Paddy Scannell (1996) calls "the doubling of place," the bed-ins were occurring in Amsterdam and Montreal as well as in the places where they were being viewed. Such a doubling of place through live witnessing (usually within the privacy of people's homes) offered the potential of reducing the sense of distance between Lennon and Ono and their audience, both physically and emotionally (Couldry and McCarthy 2004). Through television, the couple could make an intimate presentation of domesticity intended for consumption within the distant domestic spaces of their audience. Apparently understanding that media "exposure" would allow them to transgress political and ideological boundaries, John and Yoko did not have a spatial focus for their campaign—it remained mobile, a series of temporary events in various places without an operational base. Hence, viewed as a whole, their campaign for peace was placeless or, potentially, in all places at once.

Also key to the campaign, was the couple's use of tactics at various scales simultaneously. At the scale of the individual events, their choice of locations shows a deliberate attempt to harness the symbolic meanings of particular places. Representing Amsterdam as idealistic and youthful and Canada as liberal and progressive, choosing the sacred place of Coventry Cathedral, buying billboard space in global cities, appearing to "expose" their private selves to the world from their bedrooms—all of these examples show an attention to dominant or emerging cultural discourses of places. The peace campaign can also be viewed as an attempt to transgress the social space of pop culture celebrities. The media coverage of the campaign contributed to the blurring of boundaries between spaces of entertainment and spaces of activism. Beginning with "Acorns for Peace" and the 1969 campaign, through their affiliations with political radicals in the United States from the early 1970s, John and Yoko attempted to recast their subject positions as cultural signifiers—to re-place themselves on the cultural landscape.

My intent here is not to imply that what they did during their campaign for peace was especially "important" in an instrumental sense—certainly, their work was uneven, their statements varied in coherence, and one easily can question the purity of their motives. Whatever the case, the impact of Lennon and Ono on the cultural/political landscape between 1968 and 1972 remains ambiguous and a matter of continuing debate. Some authors including Ray Coleman (1992) and Jan Wiener (2000) have written on the political importance of Lennon and his peace activities and their potential for promoting political activism. Yet, clearly, using the tactics Lennon chose with Yoko Ono, such a potential, if it existed, was never realized. In similarly flattering terms, Elliott (1999, 126) contends that Lennon and Ono helped to project politics into the aesthetic and cultural domains of the times:

> The only way to out-flank traditional ... social forms is by redeeming that which politics evades and forgets ... The private realm, traditionally cast as outside and other in liberal ideology, is revealed as inherent to politics, and in this sense Lennon and Ono brought subconscious political anxieties to the fore, disclosed the personal stakes of political contestation, and underscored the peculiar significance of private protest in an era of industrialized war (Elliott 1999, 126).

However, Hodson (1987, 203) counters that there is no evidence to support this claim:

> Lennon's political analysis was vague, his strategy was nothing more than to 'keep on about peace until something happened', and the culture with which he aligned himself was itself equipped more for expressive witness to evil than effective action for good.

In an editorial titled, "Count Me Out," William F. Buckley, Jr. questions the influence and desirability of the politics espoused by Lennon and Ono. Writing in November 1990, Buckley expressed resistance to Yoko Ono's desire to have the whole world—every radio station, in every country—sing, at a given hour, the song "Imagine" as if it were a global peace anthem:

> Well, we certainly want to imagine a world in which everyone lives in peace, but, you see, that is only possible in a world in which people are willing to die for causes. There'd have been peace for heaven knows (assuming heaven existed) how long in the South, except that men were willing to die to free the slaves, and Hitler would have died maybe about the time John Lennon did, at Berchtesgaden, at age 91, happy in a Jewless Europe. There have got to be reasons that even

affected John Lennon to prefer one country over against another. I happen to know this to be the case, since a long time ago he asked me to help him get papers permitting him to live in the United States (Buckley 1990, 62).

Hence, the couple's success at re-placing themselves in relation to world events and their ability to transgress the spaces of celebrity and politics is not easily determined. In part, this is because they were proposing strategies for peace that were based on symbolic and imaginative geographies, tactics difficult to evaluate in terms of instrumental social change.

Contemporary applications and imitations

Yoko Ono has shown a persistence and loyalty to the tactics and sentiments of the original campaign for peace—and applied them in contemporary contexts. In addition to the examples I noted earlier, in 1991 John and Yoko's son Sean rewrote the lyrics of "Give Peace a Chance" and recorded the updated version with major pop stars in response to the first US intervention in Iraq (Ressner 1991). Later, 12 days after the attacks of 11 September, Ono took out a full-page advertisement in *The New York Times* that simply read: "Imagine all the people living life in peace." In a recent interview, she invoked the tone of the 1969 peace campaign in relation to the current war in Iraq, "What you have to do instead of trying to stop them with your muscles is just stand for peace – not fight for peace" (*MacLean's* 2001, 58).

Yoko has also been supportive of blatant commercial re-enactments of the bed-ins. Apparently believing that any publicity can be harnessed to promote world peace, she has provided her public support for at least one such event. For example, a bed-in was conducted by a Liverpool radio station as part of a local recognition of Lennon's 60th birthday to promote Beatles-related tourism in the city. A website advertising the event stated:

> Paying tribute to the famous "bed-in for Peace" staged 31 years ago by Lennon and his wife Yoko Ono, four couples will hop into their double beds on Friday and begin a lie-in competition to see who can stay there the longest (Beatles City 2003).

In an ironic sign of the times, the website reminded listeners that "those outside of Liverpool can watch the competition live on the Internet at www.youliverpool. com." The radio station offered the winning couple $15,700 and a trip to Montreal to stay in the same bedroom suite that John and Yoko used for their bed-in in 1969. While one can wonder what Lennon's response would be to this blatant commercialization of their peace campaign, Yoko Ono supported the Liverpool bed-in and had flowers and chocolates delivered to each bedroom with cards that read, "Imagine all the peace in the World. Love, Yoko."

The Internet has also provided serious protesters with new ways in which to apply the resistance tactics of the Bed-in for Peace. In October 2001, two New Zealand artists, Amy Berk and Andy Cox, stayed in bed while fasting and meditating on world peace for 48 hours. Their website (bed-in-for-peace 2003) explained that their bed-in was a response to the US invasion of Afghanistan following the terrorist attacks of 11 September 2001. Like the bed-in for promotion described above, this bed-in for peace utilized the World Wide Web. The website states that the event could be viewed live on the Internet "to try to connect disparate groups on-line and create a world community of peace." Berk and Cox encouraged people to "join them (in bed or not) for any or all of this 48-hour even using the free IVISIT video conferencing application."

Notably, the use of the Internet is consistent with the personalizing message of the original bed-ins that, ironically, depended on the attention of the establishment media. By using the Internet, individuals are freed from the burden of attracting media attention through celebrity, and, hence, ordinary people can have access to a worldwide audience. As Lennon and Ono had hoped that their bed-ins would do, the Internet has enabled ordinary people to access worldwide audience from within their private spaces.

Summary and implications

Although this chapter focuses solely on the Campaign for Peace, the campaign is one part of a larger story of John and Yoko's involvement in politics during the late 1960s and early 1970s. Following the peace campaign, they continued their political involvement—employing more conventional means including participating in mass protests, donating money to particular groups, and writing and recording songs about specific political situations. It has been well documented that the Nixon administration wanted to deport Lennon during this era because of his activism.

While not intending to glorify the activities of Lennon and Ono or suggest that they were especially significant or effective, in summary, I contend that the Campaign for Peace is worthy of a geographical discussion for five major reasons. First, it occurred at a time when the conventions of celebrity involvement in politics used very different approaches to space to those that John Lennon and Yoko Ono chose. Second, the couple's tactics show references to feminist, art, and critical social theory that are among the recent trends in cultural geography and other disciplines. Third, the campaign utilized the symbolic use of space and place at various scales simultaneously. Fourth, the campaign for peace foreshadowed the quasi-democratic virtual media spaces of resistance. Fifth, the influence of the campaign is both relevant to and evident in the current political climate, as the couple's spatial tactics have been imitated recently in both commercial and political contexts—both of which Yoko Ono has supported.

I observe that now, more than ever before, scholars have the analytical tools to understand the complexities and nuances of events such as the peace campaign that do not fit neatly into conventional types of "resistance." While the primary foci of geography in 1969, for example, did not make it very useful for understanding the Campaign for Peace in a socio-spatial context, much has changed. With the development of feminist geography in the 1970s, the application of critical theory and post-modernism in the 1980s, and the wide-ranging theorizations of the spatialities of popular culture and resistance evident recently, geographers and those employing geographic perspectives in other disciplines now have much to draw upon and synthesize for understanding the links between popular culture, politics, and space. Given that pop stars such as Bob Geldof, Bono and others continue to be politically active, I maintain that the study of celebrity utilization of space and place for political statements and advocacy will remain a relevant and timely area of interest for geographers and others working in the social sciences.

Despite my attempts to describe and analyze in spatial terms John and Yoko's peace events and performances, the question remains, where do we *place* the Campaign for Peace of 1969? Reflecting the range of tactics used by Lennon and Ono, the campaign appears to resist any single placement. Did it effectively offer a plan for achieving peace? No. Did it offer the possibility of doing things differently through a varied and deliberate intersection of celebrity, place, and media? Maybe. Geographer Doreen Massey (2000, 281) expresses what she sees as the current state of affairs and her frustration regarding the intertwining of resistance with commercialism:

> …with that claustrophobic constellation of family/car-ownership/supermarket shopping, having won the west, now being exported everywhere else, with— perhaps most of all—every attempt at radical otherness being so quickly commercialized and sold or used to sell … With all of this, one might well ask what are, and where are the possibilities for doing things differently?

It seems that the peace campaign of Lennon and Ono suggests one answer to the question Massey poses. It is that the radical otherness of which she speaks may be rooted in, and spatially entangled with, the features of contemporary life she lists as having won the west.

References

Ahmad, I. (2005), "Nobel Peace Laureate Wangri Maathai: Connecting Trees, Civic Education and Peace," *Social Education* 69:1, 18-22.

Beatles City (2003), <http://home.earthlink.net/exonews/xtra/beatles_city.html>, accessed June 2003.

Bed in for Peace (2003), <http://www.bed-in-for-peace/bed_in_press_release.html>, accessed June 2003.

Benjamin, W. (1986), *Reflections* (New York: Schocken Books).

Bowen, D. (1997), "'Lookin' for Margaritaville: Place and Imagination in Jimmy Buffet's Songs," *Journal of Cultural Geography* 16:2, 99-109.

Brother, N. (1956), *Men and Gardens* (New York: Alfred A. Knopf).

Buckley, W. (1990), "Count Me Out," *National Review*, 42:22, 62.

CBC Home Video (2000), *John and Yoko's Year of Peace*.

Chambers, P. (2005), "John and Yoko's Stormy Trip for Peace," *Coventry Evening Telegraph* 14 June 2005, <http://www.questia.com>, accessed 10 July 2006.

Coleman, R. (1992), *Lennon: The Definitive Biography* (New York: HarperPerennial).

Connell, J. and Gibson, C. (2003), *Sound Tracks* (London: Routledge).

Couldry, N. and McCarthy, A. (2004), *Mediaspace: Place, Scale, and Culture* (London: Routledge).

Coventry Cathedral (2003), <www.coventrycathedral.org>, accessed 3 July 2003.

Crutchlow, D. (2005), "Ono—They Don't Know Yoko! LESSON: Pupils in Trees Gift Have Never Heard of Famous Donor," *Coventry Evening Telegraph* November 2005, <http://www.questia.com>, accessed 12 July 2006.

Danto, A. (2000), "Life in Fluxus," *The Nation* 271:20, 34-6.

Domosh, M. (1999), "Those 'Gorgeous Incongruities': Polite Politics and Public Space on the Streets of Nineteenth-Century New York City," *Annals of the Association of American Geographers* 89:2, 338-53.

Dworkin, D. (1997), *Cultural Marxism in Postwar Britain: History, the New Left, and the Origins of Cultural Studies* (Durham, NC: Duke University Press).

Elliott, A. (1999), *The Mourning of John Lennon* (Berkeley: University of California Press).

Fawcett, A. (1976), *John Lennon: One Day at a Time* (New York: Grover Press).

Ford, L. (1971), "Geographical Factors in the Origin, Evolution and Diffusion of Rock and Roll Music," *Journal of Geography* 70, 455-64.

Giuliano, G. and Devi, V. (2002), *The Lost Beatles Interviews* (New York: Cooper Square Press).

Giuliano, G. and Devi, V. (1999), *Glass Onion: The Beatles in Their Own Words* (New York: Da Capo Press).

Giuliano, G. and Giuliano, V. (1998), *Things We Said Today: Conversations with the Beatles* (Holbrook: Adams Media Corporation).

Grossberg, L. (1997), *Dancing In Spite of Myself* (Durham, NC: Duke University Press).

Hodson, P. (1987), "John Lennon, Bob Geldof and Why Pop Songs Don't Change the World," in E. Thomson and D. Gutman (eds.).

Hopkins, J. (1986), *Yoko Ono* (New York: Macmillan Publishing Company).

Iglauer, E. (2001), "Pierre Trudeau: Champion of a Just Society," *Americas* 53:1, 56-8.

Inglis, I. (ed.) (2000), *The Beatles, Popular Music and Society* (London: Macmillan Press, Ltd).

Kong, L. (1997), "Popular Music in a Transnational World: The Construction of Local Identities in Singapore," *Asia Pacific Viewpoint* 38:1, 19-36.

Knopp, L. (1997), "Rings, Circles and Perverted Justice," in S. Pile and M. Keith (eds.).

Kruse, R. (2005a), "The Beatles as Place Makers: Narrated Landscapes of Liverpool, England," *Journal of Cultural Geography* 22:2, 154-62.

Kruse, R. (2005b), *A Cultural Geography of the Beatles* (Lewiston: Edwin Mellen Press).

Kruse, R. (2003), "Imagining Strawberry Fields as a Place of Pilgrimage," *Area* 35:2, 154-62.

Lennon, J. (1986), *Skywriting by Word of Mouth* (New York: Harper and Row).

Leyshon, A., Matless, D. and Revill, G. (1998), "Introduction," in A. Leyshon, D. Matless and G. Revill (eds.).

Leyshon, A., Matless, D. and Revill, G. (1998), *The Place of Music* (New York: The Guilford Press).

Marcuse, H. (1999), "Liberation from the Affluent Society," in M. Steger and N. Lind (eds.).

Massey, D. (2000), "Entanglements of Power: Reflections," in J. Sharp, P. Routledge, C. Philo and R. Paddison (eds.).

McKinney, D. (2003), *Magic Circles: The Beatles in Dream and History* (Cambridge: Harvard University Press).

Meyrowitz, J. (1985), *No Sense of Place: The Impact of Electronic Media on Social Behavior* (Oxford: Oxford University Press).

Mirer, D. (2000), 'Wishing Rooms', *Art Journal*, 59:1, 99-8.

Munroe, A. (2000), "The Spirit of YES: The Art and Life of Yoko Ono," in A. Munroe and J. Hendricks (eds.).

Munroe, A. and Hendricks, J. (eds.) (2000), *Yes Yoko Ono* (New York: Japan Society and Harry N. Abrams, Inc).

Norman, P. (1981), *Shout!* (New York: MJK Books).

Pile, S. (1997), "Introduction," in S. Pile and M. Keith (eds.).

Pile, S. and Keith, M. (eds.) (1997), *Geographies of Resistance* (London: Routledge).

Ressner, J. (1991), "Lenny Kravitz's Plea for Peace," *Rolling Stone* 598, 16.

Routledge, P. (1997), "Imagineering Resistance: Pollock Free State and the Practice of Postmodern Politics," *Transactions* NS22, 359-76.

Scannell, P. (1996), *Radio, Television and Modern Life: A Phenomenological Approach* (Oxford: Blackwell Publishers).

Sharp J., Routledge, P., Philo, C. and Paddison, R. (2000), "Entangled Humans," in J. Sharp, P. Routledge, C. Philo and R. Paddison (eds.).

Sharp J., Routledge, P., Philo, C., and Paddison, R. (eds.) (2000), *Entanglements of Power: Geographies of Domination/Resistance* (London: Routledge).

Sheff, D. (1981), *All We Are Saying: The Last Major Interview with John Lennon and Yoko Ono* (New York: St. Martin's Griffin).

Steger, M. and Lind, N. (eds.) (1999), *Violence and Its Alternatives: an Interdisciplinary Reader* (New York: St. Martin's Press).

The Journal (Newcastle, England) (2005), "We All Want the Highest Standards of Dream Bedrooms," March 19, p. 56.

Thomson, E. and Gutman, D. (eds.) (1987), *The Lennon Companion* (London: Sidgwick and Jackson).

Thrift, N. (2000), "Entanglements of Power: Shadows?" in J. Sharp, P. Routledge, C. Philo and R. Paddison (eds.).

Thrift, N. (1997), 'The Still Point: Resistance, Expressive Embodiment and Dance', in S. Pile and M. Keith (eds.).

Washington Report on Middle East Affairs (2001), "U.S. and Israeli Jews Join in 'Olive Trees for Peace'," May 2001, 96. <http://www.questia.com>, accessed 10 July 2006.

Wiener, J. (2000), "Lennon's Greatest Hits," *The Nation* 271, December 18, p. 7.

Wiener, J. (1984), *Come Together: John Lennon in His Time* (New York: Random House).

Wenner, J. (1971), *Lennon Remembers* (New York: Popular Library).

Valentine, G. (1995), "Creating Transgressive Space: The Music of kd lang," *Transactions of the Institute of British Geographers* 20, 474-85.

Chapter 3
Scales of Resistance: Billy Bragg and the Creation of Activist Spaces

Edward Jackiewicz and James Craine

Introduction

Using the music of British singer-songwriter and political activist Billy Bragg, we demonstrate how music creates and sustains multi-scalar spaces of resistance. We argue that music and its aural components may be thought of as metaphors that can be interpreted and analyzed in order to show their role in social, political, and cultural creation. We believe these aural metaphors illustrate how Bragg's music shapes spaces of activism and resistance at different scales. We believe that our discussion of Bragg can further what Connell and Gibson (2003) term the "conceptual shift within cultural geography" towards "discursively constructed arenas that are shaped by wider social relations and representative of divisions and tensions in society" (2).

By examining the role of music in the formation of *any* type of activist spaces, we believe the work of Billy Bragg can be seen as containing both larger, globalized (and more specifically, Northern) societal values and mores *and* much smaller, local (and specifically British socialist) values. Bragg's music becomes a form of resistance capable of public mobilization at a range of scales—it is a soundscape where

> sounds in themselves can occupy the site of considerable political tension. Political music, from the acoustic folk of Woody Guthrie, Bob Dylan or Billy Bragg, or the harsh sounds of punk, hip hop, to the penetrating bass thump of industrial techno, have all been associated with certain movements and ideological concerns, and employed in ways that define particular spaces and events (Connell and Gibson 2003, 220).

Bragg thereby creates an aural discourse that resists hegemonic conditions of economy and polity. We argue that Bragg's work, through the use of musical texts often juxtaposed with the spectacle of performance, does indeed "convey its meanings and values through visuals, rhythms, titles of songs and albums, the timing or releases and sometimes through the lifestyles of the performer" (Kong 1995, 450-51). Thus, much like the Sex Pistols in Mitchell's (2000) discussion of the creation of culture through the spectacle of music, Bragg provides his audiences with an understanding of, and hopefully a critical stance on, the larger issues of war-mongering and global commodification. We argue that Bragg, through his lyrics and activism, invokes and produces multi-scalar spaces of resistance, ranging from the local—such as working-class issues in the UK, to the global, such as attacks on the WTO and the global economy. Subsequently, his music is consumed in a similar manner—it resonates with listeners ranging from dockworkers in Liverpool to anti-globalization activists the world over.

Our approach to the multi-scalar economic spaces of music engages those of Jameson (1992) and Taylor (1982). Bragg's work, especially his attacks on the uneven distribution of capital across globalized economies, echoes Taylor's (1982) socially produced scalar hierarchy (cf. Marston et al. 2005). Taylor's concept consists of the micro-scale urban experience, the meso-scale nation-state and its resulting ideologies, and a macro-scale derived from the economies of globalization and we argue that Bragg uses his music to further elaborate and engage the movement of capital through space at these various scalar levels. By expanding upon Taylor, we can see that Bragg's music has a function similar to that of the film *All the President's Men* in Jameson's discussion of the multi-scalar relationship between politics and economics, especially in Bragg's ability to fuse the personal and the polemic. Relating the globalized corporate and conspiratorial nature of the business of music to Jameson's description of film as a "great corporation," we can easily arrive at the same conclusion: "If you want to say something about politics, it is by way of economic raw material" and with Bragg we have political music that *sounds* like political music and is, as Jameson indicates, "a representation that seeks to convey some conception of political relations by way of overtly political material" (1992, 67). In the twenty-first century globally connected world of commercially-produced music, the commodified culture contained within capitalist spaces is now exported much faster as more and more barriers to capitalism come down (Barnett and Cavanagh 1996). Bragg's music attacks the injustices of late capitalism becoming, as Jameson surmises, a national culture form, "in which an individual somehow confronts crime and scandal of collective dimensions and consequences" and which "cannot be transferred to the representations of global postmodernity without deep internal and structural modifications" (1992, 38). Bragg's anti-globalization stance can thus be seen as the representation of the struggle for social justice, his attempt at the internal and structural modification of the globalized economies of late capitalism. This is what Bragg strives to accomplish—he modifies everything from the spaces of performance to the spaces of distribution and consumption of his products in his struggles for social justice. We argue that it is within these globalized *and* local frameworks that Bragg agitates.

Through his musical output, Bragg attacks the oppressions brought about by capitalist social injustices. Harvey (1973, 1996) sees this struggle as an attempt on the part of the oppressed and disempowered to play a role in the politics of place construction and when people and organizations are forcibly distanced from or unable to contest social injustice, no matter what the scale, a politics of conflict and resistance is created (Wilson 1995). Tensions and controversies lead to the mobilization of the disenfranchised, a topic utilized by Bragg in many of his songs and clearly enunciated within the spaces, both physical, i.e. on stage, and metaphorical, of his performances. Indeed, the environments Bragg uses as subject matter *are* oppressive and this oppression is contested, we argue, by Bragg's particular creation of musical spaces as he aids people, through his music, in their struggle to improve their lives. As a young performer, one of his defining moments came when he supported the British miners during their strike against Conservative policies that devastated lives and communities. The oppression Bragg politicizes, in terms of social and cultural justice, can take the form of exploitation, marginalization, powerlessness or cultural imperialism. If resources are unevenly distributed throughout the environment of social power, then the reaction to these forms of oppression, no matter what the catalyst, can lead to community-based local activism directed at the planned transformation, something Bragg strives to bring about through his music. This empowerment highlights the meaning of political, social, and economic rights and brings the activity of smaller scale political action directly into the homes of the oppressed (Staeheli 1994), a task Bragg accomplishes through the release of recorded product, both auditory and visual, in the form of CDs and DVDs—the distribution of which is controlled by Bragg himself.

Moreover, we argue that Bragg's local/global dialectic provide terrains of resistance where music challenges and contests power and the (re)colonization of space by the process of capitalism. Bragg's musical resistance can be experienced as a voice of the disenfranchised in the spatial and temporal struggles against the controlling forces of capitalism. Bragg rants against the unwelcome appropriation of space, time, and daily experience by creating terrains of conflict in reaction to the globalized capitalist world, creating songs and forming strategies of resistance that intersect with class, gender, ethnicity, and sexuality in new and different ways.

Background: Victim of geography

Billy Bragg was born in Barking, Essex, England in 1957. He was initially in the short-lived band Riff Raff and then turned to a solo career, releasing his first album, *Life's a Riot,* in 1983. Subsequently, he has toured around the world many times, mostly as a solo performer. His musical influences date back to the 1976 punk rock explosion that remade English music and, to a certain extent, revitalized the global music industry. Though even his earliest work has a political slant, he became fully politicized in the early 1980s by the dockworkers' strikes and by Margaret Thatcher. Indeed, the conservative ex-Prime Minister and her party became a lasting inspiration for much of his output.

Throughout his career he has been a pro-labor advocate and has established himself as a defender of labor and labor unions and other issues to the left of the political spectrum. The political and economic assault on labor and the working class in England, which began in the 1970s and continues into the present, has been the focal point for Bragg's activism and the impetus for many of his songs. An early influence in his politico-musical evolution came when he played benefits for unemployed miners and came in contact with many politically-minded folk singers who provided inspiration to his music "... and in a way, if the miners' strike hadn't come along, I probably would never have got so ideological. I've always written about what I see, about how I see things" (Bragg cited in van der Zee 1999). He recast versions of popular protest songs including: "The Internationale," "Which Side Are You On?" and "The World Turned Upside Down;" aligning them with contemporary causes. Further inspiration came from Rock Against Racism (RAR), an anti-racism, anti-National Front faction located within the British punk rock movement from 1976 (but which later became international in scope) and he was arrested outside of the South African Embassy in Trafalgar Square during an anti-apartheid protest. He has helped to mobilize anti-globalization and anti-capitalist efforts through his music. He played benefits for the Nicaragua solidarity movement, even before he sung about the problems in Northern Ireland (Collins 1998). In 1987, he was invited to Managua by Catholic priest, poet, and Minister of Culture Ernesto Cardenal to play at the International Book Fair. Ironically, he returned to Nicaragua for the 1990 presidential elections as an international observer to witness the election that ousted the Sandinistas. This type of recognition and his commitment to sociopolitical causes has cemented Bragg's position among the political left despite the modest commercial success of his recording career.

His support of labor has him linked, for better or worse, with Britain's Labour Party. In 1985, he coined the name and helped form *Red Wedge* a collective of popular musicians opposed to Thatcher politics, who tried to politicize young people and attract them to the policies of the Labour Party. Bragg later became disillusioned with the Party when its leader, Neil Kinnock, refused to oppose military action in the first Gulf War, but Bragg's political alliance is still firmly with Labour despite their recent drift toward the political center (Collins 1998) and the contentious leadership of Tony Blair and Gordon Brown. Through his persistent musical and political activities, he established himself as a legitimate and intelligent voice for the political left and for disenfranchised social and economic groups and is often called upon by newspapers such as *The Guardian* to comment on national and local political issues:

> All my politics I've had to learn, from race in Barking, through class in the army, to sexism in later life—the hardest lesson to learn for a working-class boy. Nobody's born politically correct. When do you pick it up? When you leave school? No. I was anti-racist when I left school but at the same time I was quite nationalistic. I had yet to learn the politics and language of multiculturalism— and I learnt it listening to music (Collins 1998, 146).

Bragg's commitment to social justice also carries into the recording industry. He was a shareholder of Go! Discs, the record company responsible for releasing and distributing four of his discs. Go! Discs was a cooperative distribution operation but Bragg felt uneasy about the relationship—especially taking shares from his friends—thus, he donated his 5 percent to a trust fund to assist anyone who had worked for the company for at least two years should the company be sold off or shut down. Indeed, when Go! Discs was bought out by PolyGram Bragg's employees received a handsome supplement to their severance pay (Collins 1998, 256).

Arguably, his greatest commercial success came in 1995 when the late Woody Guthrie's daughter Nora approached Bragg to put music to lyrics that Guthrie had written more than 50 years ago but never recorded. Bragg and the US band Wilco released this music as *Mermaid Avenue* in 1999 by Elektra Records (a second *Mermaid Avenue* was also released by Elektra in 2000). In an interview, Bragg surmises the reasons for his selection by Nora:

> Woody and I have some things in common inasmuch as we both sing about unions, for instance, and there aren't many contemporary people around who sing about that. Also, I think we had similar experiences between 1930s America where the culture was as politically charged as it was in 1980s England during the Thatcher years. Although the situations were different, I think Nora recognized that her father and I came to similar conclusions about working for a better society, and perhaps the best way to do that is through organized labor (Werbe 1998).

Geography, music, and resistance: Mixing pop and politics

The geographical analysis of music provides insight into the complex linkages between place, cultural identity, and globalization—especially its manifestation within specific spaces of local activism and resistance. We agree with Connell and Gibson (2003) in that "popular music can provide opportunities for individuals or groups to assert human agency, to avert cultural homogeneity, to resist symbolically the wider social order and capitalist modes of production, and negotiate hegemonic ideologies" (272). Earlier, but useful, understandings of the discourses located within spaces of aurality are formed by Foucault (1976) as he explored the connections between speech and its association with the power relationships present in social and political constructions, by Bakhtin (1981) in his discussion of the relationship between the speaker and the listener, by Carney (1998) in a wider discussion of music geography, by Sui (2000) in his essay on aurality and its function as a metaphor for geographic space, and by Aitken and Craine (2003) in their analysis of song lyrics as discursive spaces. Smith (2000, 632) points out that researchers "can try to put some words to how bodies are experienced through music and how music is experienced through bodies" in attempts to understand the larger social and political spaces generated by soundscapes in the context of both performer and listener. Lockard (1998, xii) sees music as a "site of struggle between dominant and dissenting interests" and further states

> popular music genres and musicians, like their counterparts in literature and film, have sometimes reflected and articulated themes of cultural decolonization, neo-colonialism, cultural confrontation, disillusionment, despair, widespread inequalities, subcultural nationalism or identity, nation-building and anti-imperialism.

Musical spaces also create cultural images that represent, structure, and symbolize our surroundings; places that can be analyzed to better understand the lived experience. Aural spaces provide a site, or a micro-geography (Elwood and Martin 2000), that embody many different scales of spatial relations and meanings. These "sites" are also aural representations, or cultural signifiers that, when interpreted, reveal social attitudes and material processes where ideologies are transformed into concrete forms (Duncan and Duncan 1988). It is within these representations where Bragg's music becomes an active site of resistance, mounting an aural assault against the inequalities of capitalism and globalization.

In that sense, our study differs from other perspectives on popular music that focus on a genre of music or, more commonly, place-based music. Some examples of these geographic works are Ford's (1971) analysis of the diffusion of American rock and roll, Carney's discussions of country music (1992) and bluegrass music (1996), and Feld's (2001) discussion of the appropriation of non-Western music by Western musicians through unauthorized sampling, a topic touched upon to some degree by many authors analyzing the effects of globalization on musical propriety. Feld (2001, 198) does, however, locate discourses of globalization directly within musical space by stating music "is equally routed through the public sphere via tropes of anxiety and celebration." Similar studies include Gill's (1993) use of structuration theory to analyze music from the Pacific Northwest, Kong's (1996, 447) focus on local Singaporean music as a site of resistance in which she positions music as "a form of cultural resistance against state policies and some socio-cultural norms," Ingham et al.'s (1999) description of warehouse raves, and Blackburn and Revill's (2000) discussion of the role of music in the organization of social, economic and political spaces which explores the spatial processes of cultural production through music. Clarke (1996) explores the homogenization of pop music as a cultural product through the diffusion of MTV, and Barnet and Cavanagh (1996) discuss the inability of music that is globally distributed through traditional methods to create greater concern for the welfare of oppressed peoples and cultures. More appropriate, however, is Gill Valentine's (1995) discussion on how the music of kd lang is consumed by lesbian audiences to facilitate the production of queer space, a view particularly instructive because of Valentine's focus on the creation of identity through resistance. In addition, Smith (1997, 502) directly links spaces of music to practices of domination and empowerment by "teas[ing] out the contested terrain of music" and identifying it as "a medium through which those whose condition society tries its best not to see can begin to make themselves heard." Thus we believe Bragg's work, whether through music or activism, tends to transcend place-based consciousness for a multi-scale rendering of the colonial tendencies of Northern economies of consumption.

Through his music, Bragg creates spaces of contestation and resistance that mediate the adverse impacts of territorial and spatial confrontations, especially when the economic and social life of the world is affected in a negative manner. Resistance is often formed around local communities that seek to construct a concept of place identity appropriate to their struggle against the globalized capitalist economy. Transnational corporations step in to fill the void left by the disintegration of local life and Bragg's body of work can be seen as a reaction to the forces of uneven development foisted upon the disadvantaged and powerless by this imperialism of capital.

The production and consumption of spaces of resistance

Fans of Billy Bragg's music are also drawn to the individual himself by his commitment to social and political causes and the spaces of activism that he creates. In Valentine's (1995, 474) analysis of the music of kd lang, she identifies three "processes of consuming music" which we incorporate to discuss and analyze the music of Billy Bragg. One process is the consumption of live music performed in concert spaces. The second is the "soundscape" that provides the backdrop for our everyday lives and is often heard or overheard in public places. In this instance the consumption is not deliberate—we believe this process is less applicable to Bragg's music because his output does not have the same commercial appeal as lang's music (and thus the popularity and resultant commercial success gained through repeated "playings" of his songs on capitalist radio stations, stores, or other places of consumption). However, Bragg creates spaces through his offstage activities, a somewhat different process that nonetheless reinforces the advocacy of his music and contributes to the sustained and fervent allegiance of his fan base. The third process is the private and conscious consumption of music, which helps to bring the listener to an imaginary place, one in which Bragg's music can rectify the oppressive nature of capitalist spatialities. Whereas Valentine focuses on the consumption of lang's music, we examine both the production of music and extra-musical spaces as well—from the more private spaces of consuming a CD to the more public (and spectacularized) spaces of live performance.

Bragg's music often functions locally as anti-union resistance and, concurrently, as a larger-scale anti-globalization movement, a response to geographical spaces based on the unequal distribution of capital (or the attempt to oppress an individual or group). Thus, Bragg's soundscapes, like other aural spaces of musical resistance, are often the manifestation of some perceived inequality or injustice forced upon certain inhabitants of a space or a territory (Kong 1995; 1996). These constraints may be economic in nature (such as local wage disputes or global transfers of capital) or, in some cases, can be seen as an interruption of a social construct (such as localized anti-union activities or the globalized exploitation of female labor). In either instance, there is some form of political-economic confrontation or resistance that can, we argue, take the form of resistance-based soundscapes. In this example, the redistribution of capital perpetuates a socio-economic inequality that Bragg confronts with an enthusiastic and continuous output of both labor and product:

> Those whose lives are ruled by dogma are waiting for a sign
> The Better Dead Than Red Brigade are listening on the line
> And the liberal, with a small L cries in front of the TV
> And another demonstration passes on to history
> Peace, bread, work, and freedom is the best we can achieve
> And wearing badges is not enough in days like these
> (from "Days Like These" on *Reaching to the Converted*, 1999)

Bragg's productions, both as recorded music and as performance, can provide insight into the use of space and environment in relation to economic restructuring and the uneven distribution of capital. Bragg addresses the question of how communities are integrated into the global economy by closely relating, through his works and his labor, to issues surrounding everything from local resource management and environmental regulation to the struggles for economic stability and equality.

> Sometimes I think to myself
> Should I vote red for my class or green for our children?
> But whatever choice I make
> I will not forsake
> So you bought it all, the best your money could buy
> And I watched you sell your soul for their bright shining lie
> Where are the principles of the friend I thought I knew
> I guess you let them fade from red to blue
> (from "Red to Blue" on *William Bloke*, 1996)

In the cases of peasant and agrarian societies or working class societies forcibly undergoing a transition to a capitalist economy, Bragg's activist performative stance closely examines the global human-induced modifications of the economy. Introduced into the capitalist marketplace, these societies unwillingly become spatially marginalized in political, economic, and ecological terms as the pressure of production places severe stress on limited or diminishing human and natural resources. Bragg, in confronting these issues of social justice, alters the spatial scale within his musical production—from a singularized private person to globalized nations of people to better accommodate his confrontation with capitalist economies.

So don't expect it all to happen
In some prophesied political fashion
For people are different
And so are nations
You can borrow ideas
But you can't borrow situations
(from "North Sea Bubble" on *Don't Try This at Home*, 1991)

Bragg uses the social (and Socialist) production of space and nature to help explain capitalist pressures brought about by the globalization of economies and cultures. As societies turn to the globalized, imperialist, neocolonial economy the use of labor in the production of space contributes to the extraction of super-profits leading to the creation of marginalized social elements. As the society is changed, so is the space it occupies and, in turn, so is the culture within those spaces. By dividing his music into specific geographic regions and scales, the Marxist aspects of Bragg's output are more clearly understood and comprehended as soundscapes of resistance. Through his production of aural spaces, Bragg makes the consumer of his music aware of the uneven distribution of wealth, the redistribution of capital, and the effects of globalization on previously non-capitalist economies at various scales and differing regions and locations. Bragg challenges these capitalist-induced modifications to the environment by informing consumers of the result of the production of capitalist space and the severe and often disproportionate effects globalization has on marginalized and disenfranchised cultures. Thus, by way of Bragg's efforts, more and more space, at many different scales and places, is added to the resistance to globalized, imperialist economies, in an effort to lessen or even halt the ecological and economic impact on transitional cultures.

Outside the patient millions
Who put them into power
Expect a little more back for their taxes
Like school books, beds in hospitals
And peace in our bloody time
All they get is old men grinding axes
Who've built their private fortunes
On the things they can rely
The courts, the secret handshake
The Stock Exchange and the old school tie
For God and Queen and Country
All things they justify
Above the sound of ideologies clashing
(from "Ideology" on *Talking with the Taxman about Poetry*, 1986)

We can divide Bragg's lyrical content into categories of resistance to demonstrate the multiscalarity of his music. As previously mentioned, Bragg is well known for his attacks on global capitalism and the politics of globalization:

> Here we are, seeking out the Reds
> Trying to keep the Communists in order
> Just remember when you're sleeping in your beds
> They're only two days drive from the Texas border
> We're making the world safe for capitalism
> (from "The Marching Song of the Covert Battalions" on *The Internationale*, 1990)

Here, his target is US-led imperialism:

> They're already shipping the body bags
> Down by the Rio Grande
> But you can fight for democracy at home
> And not in some foreign land
> (from "Help Save the Youth of America" on *Talking with Taxman about Poetry*, 1986)

Here, he again contrasts the lifestyle disparity between the Global North and South:

> It may have been Camelot for Jack and Jacqueline
> But on the Che Guevara highway filling up with gasoline
> Fidel Castro's brother spies a rich lady who's crying
> Over luxury's disappointment
> So he walks over and he's trying
> To sympathise with her but he thinks that he should warn her
> That the Third World is just around the corner
> (from "Waiting for the Great Leap Forward" on *Worker's Playtime*, 1988)

A recent song attacks the US invasion of Iraq:

> Saddam killed his own people
> just like general Pinochet
> and once upon a time both these evil men
> were supported by the U.S.A.
> And whisper it, even Bin Laden
> once drank from America's cup
> just like that election down in Florida
> this shit doesn't all add up.
> It's all about the price of oil

'cause it's all about the price of oil
don't give me no shit
about blood, sweat, tears and toil
it's all about the price of oil.
(from "The Price of Oil" on *Stop the War*, 2002)

At a national scale, the economic problems and working class struggles in the 1970s United Kingdom provide the content:

At twenty-one you're on top of the scrapheap
At sixteen you were top of the class
All they taught you at school
Was how to be a good worker
The system has failed, you don't fail yourself.
(from "To Have and Have Not" on *Back to Basics*, 1987)

Here, Bragg refers to the working class struggles confronting much of 1980s England:

I kept the faith and I kept voting
Not for the iron fist but for the helping hand
For theirs is a land with a wall around it
And mine is a faith in my fellow man
Theirs is a land of hope and glory
Mine is the green field and the factory floor
Theirs are the skies all dark with bombers
And mine is the peace we know
Between the wars
(from "Between the Wars" from *Between the Wars*, 1985)

Here his anti-Thatcher feelings are made quite clear:

You Thatcherites by name, lend an ear, lend an ear
You Thatcherites by name lend an ear
You Thatcherites by name, your faults I will proclaim,
Your doctrines I must blame, you will hear, you will hear
Your doctrines I must blame, you will hear
You privatise away what is ours, what is ours
You privatise away what is ours
You privatise away and then you make us pay
We'll take it back some day, mark my words, mark my words
We'll take it back some day, mark my words
(from "Thatcherites" on *Bloke on Bloke*, 1997)

In most public discourse to label yourself a "socialist" has become somewhat of an anachronism (or worse)—Bragg, however, still uses the word with great and heartfelt conviction at a very local and personalized scale:

> I dreamed I saw a tree full of angels, up on Primrose Hill
> And I flew with them over the Great Wen till I had seen my fill
> Of such poverty and misery sure to tear my soul apart
> I've got a socialism of the heart, I've got a socialism of the heart
> (from "Upfield" on *William Bloke*, 1996)

Here, he reaffirms those Socialist convictions:

> I see no shame in putting my name
> To socialism's cause
> Nor seeking some more relevance
> Than spotlight and applause
> Neither in the name of conscience
> Nor the name of charity
> Money is put where mouths are
> In the name of solidarity
> (from "I Don't Need this Pressure Ron" on *Reaching to the Converted*, 1999)

Billy Bragg in concert: Reaching to the converted

A concert venue functions as a mostly closed space where there is typically an intimate connection between the performer(s) and the audience. Those in the audience are often familiar with the music and the persona of the performer. Within these performative spaces, a tightly knit community is formed that shares not only the appreciation of the music, but often times the socio-political perspectives of the performer. Similar to Valentine's discussion of the concert spaces created by the music of kd lang whereby traditionally heterosexual space is transformed into a lesbian space, Bragg transforms normally apolitical spaces into spaces of activism. Moreover, those in the audience who are often the minority in most public arenas, now find themselves as the majority. A concert experience is vastly different from other music listening experiences because it is visual as well as aural. The music is mediated through the body of the performer through sound, images, and movement (Valentine 1995). Many Bragg fans go to his concerts not only to hear his music, but also to listen to his rants in between songs that interlace current political issues and humor, and very often the lyrics to one of his signature songs, "Waiting for the Great Leap Forward," is rewritten to address a specific localized political event.

Bragg's concerts create a space that allows the audience to express their political views without fear of repercussion and within this arena his music represents a celebration of camaraderie and solidarity. His lack of widespread commercial popularity (often times he performs in arenas or clubs that hold anywhere between 1,000 and 2,000 people, and many times less) helps to create a sense of solidarity among the audience that continues to resonate long after the performance. Bragg also grants space at his concerts to local activist groups to help them spread their message. He also makes the audience aware of this presence thereby helping them promote their organization and spread their word to those with a similar political vision. So for Bragg, like Attali (1985), the political economy of music begins with the social distinction between nature and culture and, as explained by Revill (2000, 599), "continues through a set of processes localizing and commodifying sound, and results in the mobilization of particularized sound as a universal globalizing aesthetic, political, and economic force." The spaces created at Bragg's concerts and other performative events provide a safe haven for those with non-traditional or more alternative political perspectives who often find themselves the minority in most public arenas. Indeed, as Smith (2000, 634) states, "the way listeners themselves 'perform' has a bearing on the political, economic, and emotional spaces of music." Of course, this is not a space where political conversion is likely to occur as most see the world through similar lenses, but it does provide a place, for a few hours at least, where those from the political left and other disenfranchised groups can feel comfortable sharing and expressing their views.

Conclusion

Billy Bragg recently authored his first book *The Progressive Patriot: A Search for Belonging* which tackles the issue of national identity in contemporary England, a theme central to his 2006 album *England, Half English*. Written as his response to the London bus bombings of July 2005, Bragg questions what it means to be British in a multicultural society where such acts of person-to-person violence, often motivated by cultural differences, can occur. One of his major points in the book is that the political right has somehow controlled the notion of patriotism to promote xenophobic feelings toward immigrants and other minorities. Bragg suggests that a multicultural patriotism is indeed progressive and patriotic and is something England should strive for; a view that resonates with most countries of the West, not just England. By advocating this multicultural geopolitic through his music and now his literary efforts, Bragg has established a devoted following.

We have examined Billy Bragg's life and music in an effort to demonstrate how his music is an important geographic concept shaping cultural geographies in various ways. Popular music is one of the most widely diffused aspects of culture and yet it is a still finding its way into the discourse of the subfield. According to Smith (1994), sound is "… inseparable from the social landscape" and is thus an important area of geographical inquiry. With that in mind, we utilized the lyrics of Billy Bragg to demonstrate how he has moved from the margins of key political debates to become a legitimate voice that creates spaces of resistance that transcend global scales, moving from the local to the national to the global, sometimes in the same song.

The lyrics and actions of Billy Bragg are useful for demonstrating how music creates a space of activism and the creation of resistance to boundaries and borders, both social and political. As Smith (2000) points out, music is an economic space, while at the same time, in the case of Bragg, it is a social space or space of activism. The metaphoric spaces of musical production become the site of activism attacking capitalism, US hegemony, racism, and many other aspects of globalization. Moreover, it is a multi-scale assault on neocolonialism and the exploitation of the masses in support of the inhabitants of a certain place or space. The creation and sustenance of this space of activism is no small feat given the contemporary global political climate and Bragg continues with his convictions, even at the expense of larger commercial success. Creating music for some is not only a means to earn a living, but a vehicle to actively prosecute one's battles against the injustices of globalized capitalist economies.

References

Aitken, S. and Craine, J. (2002), "The Pornography of Despair: Lust, Desire and the Music of Matt Johnson," *ACME: An International E-Journal for Critical Geographies* 1:1.

Appadurai, A. (ed.) (2001), *Globalization* (Durham, NC: Duke University Press).

Attali, J. (1985), *Noise: The Political Economy of Music* (Minneapolis: University of Minnesota Press).

Bakhtin, M. (1981), *The Dialogical Imagination* (Austin: University of Texas Press).

Barnet R. and Cavanagh, J. (1996), "Homogenization of Global Culture," in Mander, J. and Goldsmith, E. (eds.).

Bragg, B. (2006), *The Progressive Patriot: A Search for Belonging* (London: Transworld Publishers).

—— (2003), <http://www.billybragg.co.uk/biography/index.html>, accessed May 2004.

Carney, G. (1992), "Branson: The New Mecca of Country Music," *Journal of Cultural Geography* 14:2, 17-32.

—— (1996), "Western North Carolina: Culture Hearth of Bluegrass Music," *Journal of Cultural Geography* 16: 65-87.

—— (1998), "Music Geography," *The Journal of Cultural Geography* 18: 1-10.

Clarke, T. (1996), "Mechanisms of Corporate Rule," in Mander, J. and Goldsmith, E. (eds.).

Collins, A. (1998), *Still Suitable for Miners Billy Bragg: The Official Biography* (London: Virgin Publishing).

Connell, J. and Gibson, C. (2003), *Sound Tracks: Popular Music, Identity and Place* (London: Routledge.)

Duncan, J. and Duncan, N. (1988), "(Re)reading the Landscape," *Environment and Planning D: Society and Space* 6: 117-26.

Elwood, S. and Martin, D. (2000), "'Placing' Interviews: Location and Scales of Power in Qualitative Research," *The Professional Geographer* 52:4, 649-57.

Feld, S. (2001), "A Sweet Lullaby for World Music," in Appadurai, A. (ed.).

Ford, L. (1971), "Geographic Factors in the Origin, Evolution and Diffusion of Rock and Roll Music," *Journal of Geography* 70: 455-64.

Foucault, M. (1976), *The Archaeology of Knowledge* (New York: Harper & Row).

Gill, W. (1993), "Region, Agency and Popular Music: The Northwest Sound, 1958-1966," *The Canadian Geographer* 37: 120-31.

Harvey, D. (1973), *Social Justice and the City* (Baltimore: John Hopkins University Press).

—— (1996), *Justice, Nature and the Geography of Difference* (Cambridge, MA: Blackwell).

Ingram, J., Purvis, M., and Clarke, D.B. (1999), "Hearing Places, Making Spaces: Sonorous Geographies, Ephemeral Rhythms, and the Blackburn Warehouse Parties," *Environment and Planning D: Society and Space* 17: 283-305.

Jameson, F. (1992), *The Geopolitical Aesthetic: Cinema and Space in the World System* (Bloomington: Indiana University Press).

Kong, L. (1995), "Popular Music in Geographical Analysis," *Progress in Human Geography* 19: 183-98.

—— (1996), "Popular Music in Singapore: Exploring Local Cultures, Global Resources, and Regional Identities," *Environment and Planning D: Society and Space* 14: 273-92.

Lockard, C. (1998), *Dance of Life: Popular Music and Politics in Southeast Asia* (Honolulu: University of Hawai'i Press).

Mander, J. and Goldsmith, E. (eds.) (1996), *The Case Against the Global Economy* (San Francisco: Sierra Club Books).

Marston, S., Jones, J.P., and Woodward, K. (2005), "Human Geography Without Scale," *Transactions of the Institute of British Geographers* 30: 416-32.

Mitchell, D. (2000), *Cultural Geography: A Critical Introduction* (Malden, MA: Blackwell Publishers).

Revill, G. (2000), "Music and the Politics of Sound: Nationalism, Citizenship, and Auditory Space," *Environment and Planning D: Society and Space* 18: 597-613.

Smith, S. (1997), "Beyond Geography's Visible Worlds: A Cultural Politics of Music," *Progress in Human Geography* 21:4, 502-29.

—— (2000), "Performing the (Sound)world," *Environment and Planning D: Society and Space* 18: 615-37.

Staeheli, L. (1994), "Empowering Political Struggle: Spaces and Scales of Resistance," *Political Geography* 13:5, 387-91.

Sui, D. (2000), "Visuality, Aurality and Shifting Metaphors of Geographical Thought in the Late Twentieth Century," *Annals of the Association of American Geographers* 90: 322-43.

Taylor, P. (1982), "A Materialist Framework for Political Geography," *Transactions of the Institute of British Geographers* 7: 15-34.

Valentine, G. (1995), "Creating Transgressive Space: The music of kd lang," *Transactions of the Institute British Geographers* 20: 474-85.

Uncut Magazine (2002), "Billy Bragg," April (59), 8-9.

Van Der Zee, B. (1999), "Interview with Billy Bragg," *The Guardian*, 2 September 1999.

Werbe, P. (1998), "Interview with Billy Bragg," *Detroit Metro Times*, 2 September 1998.

Wilson, D. (1995), "Urban Conflict Politics and the Materialist-Poststructuralist Gaze," *Urban Geography* 16: 8, 734-42.

Discography

Bragg, Billy (1983), *Life's a Riot* Compact Disc. Yep Rock Records.

Bragg, Billy (1985), *Between the Wars* LP. Go! Discs.

Bragg, Billy (1986), *Talking with the Taxman about Poetry* Compact Disc. Yep Rock Records.

Bragg, Billy (1987), *Back to Basics* Compact Disc. Elektra Records.

Bragg, Billy (1988), *Worker's Playtime* Compact Disc. Elektra Records.

Bragg, Billy (1990), *The Internationale* Compact Disc/EP. Elektra Records.

Bragg, Billy (1991), *Don't Try This at Home* Compact Disc. Elektra Records.

Bragg, Billy (1996), *William Bloke* Compact Disc. Yep Rock Records.

Bragg, Billy (1997), *Bloke on Bloke*, Compact Disc. EP. Cooking Vinyl Records.

Bragg, Billy (1999), *Reaching to the Converted* Compact Disc. Rhino Records.

Bragg, Billy (2002), *Stop the War* Compact Disc/Compilation. Shock Records.

Bragg, Billy (2006), *England, Half English* Compact Disc. Yep Rock Records.

Bragg, Billy and Wilco (1999), *Mermaid Avenue* Compact Disc. Elektra Records.

Bragg, Billy and Wilco (2000), *Mermaid Avenue, Vol. II* Compact Disc. Elektra Records.

Part II
Tourism and Landscapes of Music

People travel to places to experience a particular kind of music in its proper context or to experience a meaningful personal connection with musical landscapes. In Derek Alderman's "Writing on the Graceland Wall: On the Importance of Authorship in Pilgrimage Landscapes," the connection between spirituality, place, and music is strong. Certainly, the spiritual dimension is not religious in a narrow sense, but travels to Elvis's Graceland home in Memphis, Tennessee are nevertheless acts of pilgrimage for fans: tribute is paid to a revered figure. In popular culture, not only music fans are known to engage in this behavior, but also sports fans and others express reverence for special places using vernacular terms such as "meccas" or "sacred ground." In this case, Alderman focuses on one site within the larger pilgrimage landscape of Graceland; many visitors write about their relationship with Elvis on a wall surrounding the compound. What could be more appropriate when studying a pilgrimage site in popular culture than seeking the meaning of this spatial practice directly in the written record left behind by the "pilgrims?"

"Writing on the Graceland Wall" has two objectives. First, it offers a conceptual framework for interpreting Graceland and the specific cultural practice of visitors who make inscriptions on the wall outside the mansion. Graceland plays an important role in both the official and vernacular construction of Elvis as a sacred figure in popular culture. While the estate's management attempts to fix and control the official public image of Presley, the inscriptions add another dimension to the officially sanctioned, and perhaps antiseptic, perspective of Elvis. By focusing on the inscription process, this chapter emphasizes the idea of Graceland visitors as authors or creators of Elvis's memory. This authorship includes the continuous rewriting of the symbolic meaning of Graceland, prompting us to recognize the place-defining potential of music fans as a cultural group. This approach is grounded in the belief that pilgrimage landscapes are the visible articulation of competing discourses about religious heritage.

Secondly, the chapter suggests ways of analyzing the writings found on the Graceland wall and uses a published collection of these inscriptions to document how visitors interpret and construct Elvis's image. Alderman uses discourse analysis to organize the fans' writings. While discourse analysis acknowledges that multiple ways of reading texts and discovering truths are possible, the author identifies four overarching themes: Elvis as American Dream, Elvis as Food and Consumption, Elvis as Sex and Romance, and Elvis as Family. These discourses reveal how Elvis is made socially important and meaningful to visitors. This chapter advances not only our understanding of Graceland as a "sacred" place in North American music geography, but also the larger significance of memorials and monuments devoted to popular music celebrities and icons (see also Kruse 2003).

John Connell and Chris Gibson's "Ambient Australia: Music, Meditation, and Tourist Places" transects first and foremost with Alderman's chapter, but there are also connections with several themes elsewhere in this book. Connell and Gibson bring attention to a genre of music that is not frequently approached in critical analysis. Much like contemporary Christian music in John Lindenbaum's later chapter, ambient music does not generally receive critical acclaim in the music press. The content may not be seen as rich with opportunities for analysis either. For example, one aspect of ambient music that makes it different is the sparse use of lyrics. Because of the relative absence of lyrics, an emphasis is placed on the aural and the meaning of sounds rather than the meaning of lyrics. The cultural significance of soundscapes is a topic that has recently generated a growing body of research—see Anderson, Morton, and Revill (2005). Ambient music seeks influence in nature, while human voices, in a literal sense, are usually excluded. The traces of authorship in ambient music are erased to give the impression of a seamless connection with the natural.

As ambient music seeks inspiration in non-western musical tradition, Connell and Gibson's chapter is different than most chapters in this book. Certainly, reggae, as discussed later by Sarah Daynes, exists outside the mainstream of western popular music, but reggae can also claim indigenous ownership as it is performed by Jamaican musicians and is a leading vehicle for a distinct local, albeit globally disseminated, cultural expression. Ambient music coopts and commodifies an Aboriginal tradition: its production and performance generally does not include Aborigines, and it is produced for a global audience rather than for an indigenous one. Because the production process is in the hands of commercial interests, albeit on a relatively small-scale, and the music is produced for tourist markets, ambient music exists in the realm of popular music.

Connell and Gibson are not, however, limited to representations of idealized Australian landscapes (see also Diamond [2005] for a critique of the naturalness of Australia). Their study is positioned at the confluence of real and imaginary places, where representational and material research approaches meet. A central argument in their chapter is that music can materially transform places. Such place transformations can either be urban, such as districts in the two Australian towns of Alice Springs and Byron Bay that cater to tourists with an interest in ambient music, or economic, where the benefits of music accrue to certain people and places rather than others. Here, the authors build on their existing research on tourism and geography (2005) and how this combination has the capacity to produce new landscapes.

References

Anderson, B., Morton, F., and Revill, G. (2005), "Practices of Music and Sound," *Social & Cultural Geography* 6:5, 639-44.

Diamond, J. (2005), *Collapse: How Societies Choose to Fail or Succeed* (New York: Viking).

Gibson, C. and Connell, J. (2005), *Music and Tourism: On the Road Again* (Clevedon, UK: Channel View).

Kruse, R. (2003), "Imagining Strawberry Fields as a Place of Pilgrimage," *Area* 35:2, 154-62.

Chapter 4

Writing on the Graceland Wall: On the Importance of Authorship in Pilgrimage Landscapes[1]

Derek H. Alderman

Perhaps no other figure is more identified with modern popular music in English than singer Elvis Presley. Since his death in 1977, Elvis has transcended the boundaries of rock n' roll celebrity to become a ubiquitous image and icon in contemporary culture. According to Rodman (1996, 1), Elvis's image appears in

> songs, movies, television shows, advertisements, newspapers, magazines, comic strips, comic books, greeting cards, trading cards, T-shirts, poems, plays, short stories, novels, children's books, academic journals, university courses, art exhibits, home computer software, cookbooks, political campaigns, postage stamps, and innumerable other corners of the cultural terrain.

Presley is an omnipresent icon for many reasons, some of which we will never understand. Elvis's fans are an important but often under-analyzed factor behind his continued popularity. They play an active role in "preserving and controlling Elvis's history" and "revising and redeeming his historical memory (Doss 1999, 64)." Devotion to the King runs so deep and strong that public remembrance of Elvis Presley has taken on religious-like meaning to certain social groups inside and outside the United States. The Elvis phenomenon signals the extent to which music fandom can rival, if not exist alongside conventional notions of sacredness.

Fans connect with Elvis through a variety of cultural practices, from impersonation to the collecting of images and objects related to the entertainer. Arguably, the best known and most public of these practices is the pilgrimage to Graceland—Elvis's last residence and final resting-place in Memphis, Tennessee. Christine King (1993) has gone as far as to compare the Graceland estate to the shrines of medieval European saints, noting similarities in the sense of validation

---

1   This chapter is a lightly modified version of Alderman, D.H. (2002), "Writing on the Graceland Wall: On the Importance of Authorship in Pilgrimage Landscapes." *Tourism Recreation Research* 27:2, 27-34. It is reprinted with the permission of the publisher of *Tourism Recreation Research*.

and devotion felt by pilgrims. Approximately 700,000 visitors tour the mansion each year, with daily admission reaching as much as 5,000 people in the busy summer months (*Los Angeles Times* 1997). As Coffey (1997, 289) contended: "Graceland may be the second most famous house in America, next to the White House." When faced with the choice, 18 percent of registered voters surveyed by Fox News in 1997 actually said they would rather visit Graceland than the White House (Fox News 1997).

While now a popular destination for avid fans and curious observers across the country and the globe, Graceland originated out of Presley's need for sanctuary. Arguably, there would have never been a Graceland if not for the first home bought by Elvis on 1034 Audubon Drive. The house in Audubon Park, a quiet, upper middle-class neighborhood in Memphis, became a center of fan commotion with the arrival of the Presley family, a fact that tried the patience of Elvis's mother Gladys and horrified neighbors. According to Memphis historians Cindy Hazen and Mike Freeman (1997, 89), these neighbors "banded together and offered to buy the Presley home until they discovered that Elvis was the only homeowner without a mortgage." In March of 1957, less than a year after buying the house on Audubon Drive, the Presleys purchased the more spacious and private Graceland mansion. Upon moving in, Elvis constructed a fieldstone wall around the estate to shield himself and his family from the traffic of admirers and observers. Noted Elvis scholar Vernon Chadwick described the important symbolic role that this wall has come to play in the Elvis commemorative movement and the journey to Graceland.

> Ironically, the wall and gates that were meant to secure Elvis's privacy soon became a magnet for the 24-hour vigil maintained by legions of fans, some of whose scribblings, like initials of ancient tourists etched in the pyramids of Gezer, now can be read by all. While Elvis was alive, but increasingly after his death, visitors to Graceland found the wall a tempting space for a variety of communiqués: personal greetings, jokes, prayers, poems, love letters, votive offerings. What normally would be considered acts of vandalism were in fact acts of devotion by some of the politest and best-intentioned fans in the world (quoted in Wright 1996, 7).

This chapter has two purposes. First, it offers a conceptual framework for interpreting Graceland and the specific cultural practice of visitors making inscriptions on the wall outside the mansion. Graceland is a pilgrimage landscape devoted to "cultural religious heritage." It plays an important role in both the official and vernacular construction of Elvis as a sacred figure in popular culture. While the estate's management attempts to fix and control the official public image of Presley, writing on the Graceland wall represents vernacular and more diverse ways of sanctifying the singer. By focusing on the inscription process, I emphasize the idea of Graceland visitors as "authors" or creators of Elvis's memory and religiosity. This authorship is not limited to Elvis's image but includes the constant rewriting of Graceland materially and symbolically, prompting us to recognize the place-defining potential of music fans as a cultural group.

Second, the study suggests a way of analyzing the writings found on the Graceland wall and uses a published collection of these inscriptions to document a few of the ways in which visitors interpret and construct Elvis's image. This approach is grounded in the belief that pilgrimage landscapes are the visible articulation of multiple, sometimes competing discourses about religious heritage. Several discourses flow through inscriptions on the Graceland wall, thus revealing the versatile way in which Elvis is made socially important and meaningful to visitors. This study hopes to advance not only our understanding of Graceland as an important, perhaps "sacred," place in North American music geography, but also the larger landscape significance of memorials and monuments devoted to popular music celebrities and icons. Indeed, Graceland was officially designated as a National Historic Landmark in 2006 (Gibson and Connell 2007).

Graceland as a pilgrimage landscape: Visitors as authors

The historian of religion Juan Campo (1998) noted the popularity and diversity of pilgrimage landscapes in contemporary America. He introduced a typology for categorizing pilgrimages and, specifically, the journey to Graceland. First, there are pilgrimages associated with *organized religions*, such as the Catholic pilgrimage to the National Shrine of St. Jude in Chicago or the Mormon pilgrimage to Temple Square in Salt Lake City. Second, there are pilgrimages associated with American *civil religion* or the "interconnection of God and country, commemoration of heroes … and martyrs, and the attribution of patriotic significance to the natural landscape" (Campo 1998, 48). Such patriotic journeys take people to the Gettysburg National Military Park, Mount Rushmore, the United States Holocaust Museum, and the Martin Luther King, Jr. National Historic Site.

Third and most important to the subject at hand, there are pilgrimage landscapes associated with *cultural religion*, the sanctification of figures and themes from popular culture and mass media, including the music industry. These pilgrimage landscapes are often sites for the production of nostalgia and the commodification of culture. According to Campo (1998), Disney theme parks and Graceland are the two best examples of pilgrimage landscapes associated with American cultural religion. Campo's designation of Graceland as a cultural pilgrimage landscape recognizes the sacred-like status of Presley while avoiding the obvious pitfalls of viewing his memorialization in the same way as conventional faiths. Indeed, from the perspective of many fans, the commemoration of Elvis does not necessarily replace or conflict with other forms of religious heritage. Visitors to Graceland often celebrate Elvis's memory by referring to and embracing the entertainer's dedication to the "organized religion" of Christianity as well as his commitment— through military service—to the "civil religion" of American patriotism (Olson and Crase 1990).

Campo (1998) used the word pilgrimage landscape, as I do here, to stress the interrelationships between people and place. No place is intrinsically sacred. Pilgrimages, and their attendant landscapes, are social constructions—a perspective that recognizes the role played by "humans in their creation, appropriation, organization, and representation" (Campo 1998, 42). Pilgrimage landscapes do not simply emerge but undergo what Seaton (1999) called "sacralization." Sacralization is a sequential process by which tourism attractions are "marked" as meaningful, quasi-religious shrines. Graceland is not a cultural religious landscape simply because it was once the home of Elvis Presley. Rather, its holiness stems from the behaviors, attitudes, and actions displayed by visitors (Rodman 1996). The cultural significance of Graceland, according to Gilbert Rodman (1996, 103), is that it serves as "a physical point of articulation, where a global community of Elvis's fans regularly congregate and acquire a true sense of themselves as self-defined community." As Rodman also pointed out, however, Graceland represents a point of articulation for non-fans as well, where visitors can recognize and respond—sometimes negatively—to the canonization of Elvis and the existence of a community of fans. Graceland functions as a site for controlling, redefining, and perhaps even challenging the deification of Elvis. This chapter adopts such a perspective as it attempts to analyze how visitors use the wall outside of Graceland as a text for marking and interpreting the cultural religious importance of Elvis and the journey to Memphis.

As a pilgrimage landscape, Graceland is a site for constructing both official and vernacular images of Elvis. This distinction between official and vernacular cultural expression is drawn from the work of Bodnar (1992), known for his analysis of public commemoration in America, and J.B. Jackson (1984), known for his analysis of America's vernacular landscapes. Both authors suggested that official expressions of culture are intended to maintain social order, present a unified vision of reality, and remind citizens of their obligations and history. In contrast, vernacular constructions of culture result from the changing expressions, values, and decisions of individuals at the grassroots level. While less stable than their official counterparts, vernacular landscapes provide a mode of expression to social actors and groups not normally heard from or seen.

As the headquarters of Elvis Presley Enterprises, Graceland is deeply involved in constructing and marketing an official image of the King through the selling of copyrighted memorabilia in gift shops and the leading of tours through the estate. Visitors on these tours wear headsets playing recorded narration. There are opportunities for visitors to tour and internalize at their own pace and sequence. From this perspective, however, visiting Graceland is very much about being a *consumer* and allowing the estate's management to script and regulate much of the sacralization process. The consumer metaphor is perhaps a good word choice given that 60 percent of visitors to Graceland buy something at the gift shops before leaving (Arthur 1989). As Bodnar (1992) suggested, the goal of official commemoration is to construct and present a unified narrative about the past. Graceland actively participates in not only sacrilizing the image and life story of Elvis but also "sanitizing" it. For example, the estate's official line of tours, museum displays, and video documentaries make no mention of the racial and sexual controversies that surrounded Elvis's rise to stardom, the prescription drug abuse that led to his death, or even the fact that the King of rock n' roll died in a bathroom (Rodman 1996).

The process of sacralizing a tourist attraction such as Graceland is not just controlled by the sponsors of the site. The changing motives and perspectives of visitors and pilgrims also shape it (Seaton 1999). As Jackson (1984) and Bodnar (1992) argued, landscape change can be vernacular, or reflect the diverse interests and views of ordinary people. Two vernacular memorial practices stand out at Graceland: the leaving of personal items and offerings at Elvis's grave site, and more important to the subject hand, the inscribing of visitors' names, hometowns, and messages on the wall in front of the estate. The line between official and vernacular is perhaps not as distinct as Jackson and Bodnar suggest. Writing on the Graceland wall is not entirely unofficial. Graceland's management has built a drive-up lane adjacent to the wall to make writing on the wall safer and more accessible. They also pressure wash the wall when profane or irrelevant inscriptions appear (Rodman 1996; Wright 1996). Moreover, many of the devoted fans who visit Graceland and write on the wall are members of fan clubs that carry on an official relationship with Elvis Presley Enterprises, although not all of these clubs are on friendly relations with the estate (Doss 1999).

Although not independent of official organizers, inscriptions on the Graceland wall do represent a departure from the single, unified historical narrative presented by Elvis Presley Enterprises. They reveal the plurality of ways in which visitors remember and meaningfully connect with Elvis as part of their cultural religious heritage. Visitors frequently have their photographs taken near the wall, signifying the importance of writing and reading about Elvis as a ritual performance at Graceland. From this perspective, making the pilgrimage to Graceland is about being an *author* of Elvis's memory rather than simply a consumer of it. Viewing Graceland visitors as authors of Elvis's memory is consistent with recent research by Doss (1999). She stressed the agency and power of fans in sustaining and shaping the popularity of Elvis's image. On this point, Doss (1999: 31) wrote: "Fans do not simply derive meaning from Elvis' image, but actually 'make' Elvis in dramatic and deeply emotional ways." Of course, not everyone who inscribes a message on the Graceland wall is a devoted fan of Elvis Presley. In fact, some inscriptions are critical of Elvis's legacy. This fact does not lessen the importance of the entertainer as a cultural icon. It illustrates, as Doss (1999, 16) contended, that "Elvis' image is a model of hybridity: different images of Elvis, or in other words, different Elvises hold different meanings to different viewers." On a more general level, it signifies that the authorship of pilgrimage landscapes is not restricted to one category of visitor or pilgrim.

Toward a discourse analysis of pilgrimage landscapes

If the Graceland wall is a public text for writing about and "making" Elvis's image, then how should we read or analyze it and other landscapes like it. Schein (1997) has provided a useful framework for analyzing cultural landscapes, although he has not addressed the specific issues of music heritage and pilgrimage. He suggested that landscapes are the product of "countless individual decisions," each of which is embedded within a discourse. A discourse is a way of talking about, writing, or otherwise representing the world to make it intelligible and meaningful. Discourses are more than simply ideas and language. They are shared, common sense assumptions that shape our realities and hence the landscapes we construct (Barnes and Duncan 1992). Schein (1997, 663) characterized a landscape as the materialization of several discourses and pointed to "the possibility that the cultural landscape can itself capture different, even competing, sets of meaning, or independent, thematic networks of knowledge."

Pilgrimage landscapes can be read or interpreted as the tangible product of several discourses about the meaning of religious heritage. Galbraith (2000) advocated a similar idea when examining the experiences of young Poles on a walking pilgrimage to the Black Madonna of Czestochowa. She viewed pilgrimages as occasions for both constituting and contesting religious communities. In common with this chapter, Galbraith (2000) stressed the vernacular experiences and actions of individual pilgrims over official and institutional organizers. She characterized pilgrimage as an "arena for competing religious and secular discourses," in which pilgrims express a wide range of motivations for making the journey and, in some instances, actually challenge the official goals and interpretations of the pilgrimage (Galbraith 2000, 65).

Although not in the context of Elvis, Kruse (2005) has studied pilgrimage landscapes devoted to popular music figures in terms of discourse. In examining the Beatles, he has documented the wide range of cultural discourses that surround the band and the places associated with them. Many of these discourses go beyond the subject of music and deal with social class, national identity, race, and masculinity. For example, Strawberry Fields—the memorial to John Lennon in Central Park, New York—is more than just a place for visitors to remember the ex-Beatle but also "a spatial focus for discourses relating to world peace, leftist politics, and existential hope" (Kruse 2003, 161). Kruse also noted that music-related places of memory can become centers of tension between "authorized" and "unauthorized" commemorative practices, such as when Lennon fans contested attempts by local government authorities to impose a curfew on an annual all-night vigil held at Strawberry Fields. As Kruse (2005) found with the Beatles, I suggest that Elvis can be viewed as a "plural text," signifying more than just his music and stardom, and that Graceland's wall is a vehicle for circulating several discourses about Elvis as a culturally important figure, some of which run counter to the official tourism narrative maintained and presented by Elvis Presley Enterprises.

As suggested by scholars such as Rodman (1994), Chadwick (1997), and Doss (1999), Elvis's cultural importance is rooted in the very fact that his image is multifaceted. Elvis continues to have an active posthumous career, according to Rodman (1994), because his life story dovetails with a series of important cultural mythologies or discourses about race, gender, and class. Images of Elvis can be contradictory, such as seeing Elvis as a musical/racial integrator versus seeing him as an appropriator of black music and heritage. The imagery of Elvis has found its way into "struggles between high culture and low culture, youthfulness and adulthood, the country and the city, rebellion and conformity" (Rodman 1994, 458). References to Elvis have surfaced in the most unlikely of discussions, such as about the Gulf War, the fall of Communism, and abortion (Rodman 1994). Doss (1999) also noted the versatility of Elvis's image and how he can exist at many different levels to many different people depending on their personal and social identity. She further contended that the popularity of Elvis's image would die if left solely in the hands of his "official" sponsors. "But if Elvis remains elastic, so that fans and others can continue to look at him and talk about him and remake him … then Elvis may well survive into the twenty-first century as America's premiere icon" (Doss 1999, 259). According to Chadwick (1997), the Elvis text—that is, how Elvis's image is made part of the language of modern culture—is the product of several discourses.

Analyzing inscriptions on the Graceland wall

What sort of Elvis discourses can we find flowing through the inscriptions found on the Graceland wall? To answer this question, I drew upon the work of Daniel Wright (1996), who transcribed and published hundreds of messages from the wall into a popular book entitled *Dear Elvis: Graffiti from Graceland*. While offering little in the way of analysis, Wright's work presents the researcher with a useful database of cultural expressions. Wright does not disclose how he chose which inscriptions to publish, other than to say he selected "the best." However, the emphasis here is not on obtaining a representative sample of writings, if this is even possible. As pointed out by Berg and Kearns (1996), discourse analysis is characterized by an absence of: (1) a single set of codified procedures; (2) a single way of reading texts; and (3) a single truth to be discovered. However, this type of analysis requires close and skillful reading and interpretation.

Wright's book *Dear Elvis* contains approximately 320 messages collected from the Graceland wall in the early 1990s. This chapter presents a small sample of these inscriptions with the hope of illustrating the potential value of conducting a much larger and more empirically-based study, one that would combine the collecting of wall messages with the observing and interviewing of Graceland visitors. Based on an analysis of wall inscriptions collected by Wright (1996), visitors employ and emotionally connect with several different discourses about Elvis—Elvis as the American Dream, Elvis as Food and Consumption, Elvis as Sex and Romance, and Elvis as Family. These are not the only themes evident but represent some of the most dominant and interesting ones that Wright found in his study of the Graceland wall. The inscriptions examined here illustrate how the sacralization or marking of Graceland as a meaningful shrine is not just carried out by the estate's management but by ordinary, vernacular interests.

Elvis as American Dream

As Doss (1999) and Rodman (1994) found in their studies, Elvis is often represented as the embodiment of the American Dream of social mobility. Fans embrace him as someone who made tremendous gains in wealth and status without forgetting his working class roots. While this discourse does not dominate the messages published in Wright's collection, some visitors to Graceland do define Elvis's image in terms of how far he had come economically and socially. For instance, one inscription read: "You came to us, a poor boy from Tupelo, and you gave us all you had! You rocked the world! You will always be the King!" Some fans dwell less on Elvis's humble beginnings and more on the conspicuous wealth and success exhibited at Graceland. For instance, a visitor named Ben left this inscription: "Now that I've been to Graceland and seen it, I have the utmost respect for you." In his message on the wall, Jeff wrote: "Elvis, I loved that Jungle Room!" In a direct reference to Graceland's plush décor, another inscription read, "E. Thanks for carpeting the ceilings of our hearts!" One brave visitor to the wall requested, "Elvis, lend us some money." Not all visitors are in complete awe of Graceland, however. Connie left this observation behind: "You had a weird decorator. Thank God he didn't write music!" Elvis's penchant for buying automobiles was on the mind of Mary Lou as she contemplated his premature death and wrote: "You wouldn't have liked the way Caddies look today, anyway." One visitor expressed a critical perspective on the usually euphoric American Dream image of Elvis "Too much, too soon, Elvis." This inscription is useful in illustrating how the Graceland wall is not simply a medium for fans to express their undying devotion to Elvis. Rather, it is also a place for some visitors to conduct a sober dialogue about the ultimate meaning of the singer's life and image.

Elvis as Food and Consumption

Elvis's connection with food and eating is another discourse found flowing through the wall messages collected by Wright. The official Graceland tour openly discusses the King's love of certain types of food such as fried peanut butter and banana sandwiches. This official narrative, however, never discusses gluttony or the weight problems Elvis experienced in the last years of his life. In fact, his weight gain is blamed on chronic back problems in the official Graceland tour. Fans and other visitors to the wall do not always present such a sanitized image. One inscription read, for instance, "Turkey loves Elvis and vice-versa I bet!" Other messages proclaimed, "In the name of Elvis, I eat cheese," and "I saw Elvis at Denny's." "Elvis: How about a tennis lesson?" was perhaps a biting reference to the entertainer's exercise regiment (or lack thereof). Most devoted fans would take issue with these characterizations, arguing that Elvis lived a clean, athletic lifestyle through almost all of his years. Nevertheless, the discourse found within these inscriptions presents a more varied and richer understanding of how Graceland visitors view, identify with, and comment on Elvis's image. Wright did find inscriptions that made a more positive equation of Elvis with food. Cheryl wrote, for instance, "Have a peanut butter sandwich on me." Another inscription read, "Elvis, you're all that *and* a bag of chips." Tany had this food for thought: "Nothing better than a cheese steak, a pint, and Elvis at Cauley's." In a more puzzling and perhaps sensual food representation of Elvis, an unidentified visitor to Graceland left these words: "Elvis, you were in my dream and you were eating hot buttered popcorn." The identification of Elvis with food—while often the fuel for jokes and ridicule—deals with the fundamental issue of consumption and how one represents the body of Elvis as a cultural fact. The next section presents a very different discursive construction of Elvis's body.

Elvis as Sex and Romance

Many wall messages focus on the romantic and sexual aspects of Elvis's image. As mentioned earlier, this is a discourse not frequently found in the official narratives presented by Graceland's management. Gina wrote, for instance, "Elvis, I wish I was your belt buckle for a day!" Jenny issued this invitation: "Elvis—You are the sexiest man I have ever seen. Meet me at K-Mart (in lingerie)!" Rachel posed a question perhaps on the minds of many Elvis fans: "Hey Elvis, Are you lonesome tonight?" A visitor named Linda appears to be lonesome herself. She wrote this to the King: "I'm still waiting for my one night with you! Love you forever!" Linda will have to make room for Valerie, who inscribed "Lay your head on my pillow. Lay your warm and tender body next to mine." Anna described Elvis as "one handsome hound dog" while Cammi proclaimed, "Elvis, I lust you hard." Of course, the sexualization of Elvis's image may not appeal to all who visit Graceland. Indeed, one message read: "Elvis, Elvis, Let me be. Keep your pelvis far from me!"

Connecting to the memory of Elvis through the discourse of romance and sex is not limited to women visitors, however. Wright transcribed the following bizarre message from the Graceland wall: "Elvis, I'm bearing your ghost child, and I'm confused because I'm a man!" Upon visiting Graceland, Dave issued this compliment: "You had great taste in women. Priscilla is a babe." According to one visitor, Elvis's romantic skills are believed to carry on into the afterlife: "Please show Jackie O. a good time!" Romance was on the mind of Chris when he asked: "Yo Elvis! How about getting me and Muriel together?" As this message from

Chris shows, fans connect with the memory of Elvis at different levels and through different discourses depending on their personal identity and needs.

Elvis as Family

The discourse of family frequently surfaces as visitors to Graceland emotionally connect with the memory of Elvis. The entertainer is often remembered as part of a loving family which he supported and showered with gifts. Embracing this idea, a visitor addressed the following message to the King's parents: "Dear Gladys & Vernon—You done good!" Elvis is represented as a family man even in death. "There is only one King, and I'm sure he is having fun with his mother and father in Heaven," wrote one fan. Some fans, however, have used the wall to comment on the surviving members of the Presley family. For example, Mary pleaded: "Hey Elvis, Lisa Marie made a big mistake! Come back and straighten her out!" Stephanie added: "Elvis, How 'bout that son-in-law?" These messages, of course, refer to the short-lived marriage between Michael Jackson and Elvis's daughter Lisa Marie Presley.

Visitors also invoke the discourse of family as they recount what Elvis meant to their own loved ones. For example, Michelle wrote: "Elvis, I have grown to love you because of my Mom." Tracy showed her appreciation by leaving this message: "Thank you for making my Mom as happy as Roger has made me!" Unfortunately, Rima could not say the same: "My momma doesn't like your music, but I'm kosher." While representing two different relationships between Elvis and mothers, both messages affirmed the importance of family in contextualizing the experience of Graceland. In some cases, visitors use the wall as a medium for honoring not only Elvis but also a lost family member. Debbie wrote the following message to her father: "Dad, you are now in heaven with the King of rock-n-roll! You and Elvis will stay in my heart and mind forever." Another visitor requested that Elvis "Say Hi to Dad for me. You're both in my heart." Flo provided one of the saddest family-related inscriptions, "Elvis, my husband Mac wanted to come with me, but now he's in heaven with you." Fans identity with Elvis's devotion to family and claim him as part of their own families and family histories. This was perhaps no more evident than in a message left by Kathy: "Elvis, we love you so much that we named our little boy Chadwik after you in Blue Hawaii. We miss you."

Concluding remarks

Studying the journey to Graceland can be a difficult project for two major reasons. First, there is uncertainty about who exactly travels to Graceland, an uncertainty perpetuated by the proprietary manner in which Elvis Presley Enterprises protects visitor information for marketing purposes. Second and more importantly, scholars and the lay public have typically viewed the memorialization of Elvis as a strange, almost laughable part of popular culture. Doing so has limited our ability to study Graceland as a legitimate pilgrimage landscape and an important site of music heritage and memory. This chapter has attempted to place the study of Graceland and its visitors within a more critical light.

In summary, I have suggested that visitors to Graceland—rather than simply being passive consumers of Elvis's memory and religiosity—actually play an important role in representing the singer and his home as socially important. The memorialization of Elvis by ordinary people occurs at many different levels and at a variety of locations across the globe. However, an analysis of the Graceland wall allows us to witness the literal rewriting of Elvis's image beyond the immediate "official" control of Elvis Presley Enterprises. The reading or interpretation of these writings has been facilitated, conceptually, by viewing pilgrimage landscapes as the product of multiple discourses about the meaning of religious heritage. While sometimes appearing odd or silly, the messages found on Graceland's wall represent the convergence of different and, at times, competing discourses about Elvis and the meaning of his image. Whether one is a fan, a curious observer, or an indifferent tourist, no single image of Elvis is embraced at Graceland. Rather, visitors represent and identify with many different Elvises. To some, Elvis is sex and romance. To others, he is the embodiment of success, wealth, and consumption (even over-consumption). For many ardent fans, the prevailing discourse is one of family.

Clearly, more discourses can be read from the inscriptions found on Graceland's fieldstone wall. The author of one wall message perhaps stated it best, "Elvis, I love you for 100,000 reasons." Graceland—like any pilgrimage landscape— can attract a complex range of visitors. Visitors can include those who come for spiritual reflection, those (like myself) who seek knowledge about the social and cultural aspects of Graceland, those who come for fun or spectacle, and even those who come for all these reasons. As suggested by Jackowski and Smith (1992) and Nolan and Nolan (1992), the categories of pilgrim and tourist conceal a rich diversity of motivations, expectations, and experiences. While fans who love Elvis often write the Graceland wall, those who are critical of the phenomenon also construct it. One inscription stated it simply: "I preferred the Beatles." Future work should focus, more closely, on the politics of Elvis's memory and the role that Graceland plays in these struggles over the meaning and importance of the singer's image.

There are other pilgrimage landscapes where we can examine the role of fans as "authors" of Elvis's memory and religiosity. For example, there is Tupelo, Mississippi, where visitors can tour the small house in which Presley was born as well as a chapel, park, museum, and street bearing his name. Realizing the profits to be made, the city of Tupelo is promoting its connection with Elvis more aggressively than in the past. Another, more minor pilgrimage site is Lauderdale Courts—a public housing project in Memphis where Elvis lived as a teenager from 1949 to 1953 (Alderman 2008). While at "the Courts," Elvis attended high school and began performing publicly, using the laundry room in the basement and the front steps of the apartment building as practice areas. Incidentally, mobilized fans saved the apartment where Elvis grew up from demolition, thus further illustrating the need to view Elvis's faithful as important cultural and even political agents.

I would encourage others to explore the authorship and agency of visitors within all pilgrimage landscapes, whether those landscapes support traditional religions or unconventional faiths based in popular culture such as found at Graceland. The analysis of authorship should not be confined to the study of inscriptions and message writing. Rather, visitors author or construct pilgrimage landscapes through a variety of social practices. For example, something as seemingly innocent as walking can be interpreted as an important tourist and memorial performance (Edensor 2000). The Internet is another, emerging forum for exploring the authorship of pilgrims. Users can visit, for example, a virtual Graceland and post a message on the King's online wall (http://books.dreambook.com/gracewall/king. sign.html). These inscriptions, like their terrestrial counterparts in Memphis, may reveal even more ways in which the faithful make and remake Elvis's image as part of their cultural religious heritage.

References

Alderman, D.H. (2008), "The Politics of Saving the King's Courts: Why We Should Take Elvis Fans Seriously," *The Southern Quarterly* 46:1, 46-77.

Arthur, C. (1989), "Road Trip: Going to Graceland," *American Demographics* 11:5, 47-8.

Barnes, T. and Duncan, J. (eds.) (1992), *Writing Worlds: Discourse, Text and Metaphor in the Representation of Landscape* (New York, NY: Routledge).

Berg, L.D. and Kearns, R.A. (1996), "Naming as Norming: 'Race,' Gender, and the Identity Politics of Naming Places in Aotearoa/New Zealand," *Environment and Planning D: Society and Space* 14:1, 99-122.

Bodnar, R.E. (1992), *Remaking America: Public Memory, Commemoration, and Patriotism in the Twentieth Century* (Princeton, NJ: Princeton University Press).

Campo, J.E. (1998), "American Pilgrimage Landscapes," *Annals of the American Academy of Political & Social Science* 558: July, 40-56.

Chadwick, V. (ed.) (1997), *In Search of Elvis* (Boulder, CO: Westview Press).

Coffey, F. (1997), *The Complete Idiot's Guide to Elvis* (New York: Alpha Books).

Doss, E. (1999), *Elvis Culture: Fans, Faith, and Image* (Lawrence: University Press of Kansas).

Edensor, T. (2000), "Staging Tourism: Tourists as Performers," *Annals of Tourism Research* 27:2, 322-44.

Fox News. (1997), Opinion Dynamics Poll. August 11. Available at Roper Center for Public Opinion Research, University of Connecticut (question ID: USODFOX.081197, R3).

Galbraith, M. (2000), "On the Road to Czestochowa: Rhetoric and Experience on a Polish Pilgrimage," *Anthropological Quarterly* 73:2, 61-73.

Gibson, C. and Connell, J. (2007), "Music, Tourism, and the Transformation of Memphis," *Tourism Geographies* 9:2, 160-90.

Hazen, C. and Freeman, M. (1997), *Memphis: Elvis-Style* (Winston-Salem, NC: John F. Blair, Publisher).

Jackowski, A. and Smith, V.L. (1992), "Polish Pilgrim-Tourists," *Annals of Tourism Research* 19:1, 92-106.

Jackson, J.B. (1984), *Discovering the Vernacular Landscape* (New Haven, CT: Yale University Press).

King, C. (1994), "His Truth Goes Marching On: Elvis Presley and the Pilgrimage to Graceland," in I. Reader and T. Walter (eds.).

Kruse II, R.J. (2003), "Imagining Strawberry Fields as a Place of Pilgrimage," *Area* 35:2, 154-62.

—— (2005), *A Cultural Geography of the Beatles* (Lewiston, New York: The Edwin Mellen Press).

Los Angeles Times (1997), "700,000 Graceland Visitors a Year Can't Be Wrong," 3 August: Calendar Section, p. 8.

Nolan, M.L. and Nolan, S. (1992), "Religious Sites as Tourism Attractions in Europe," *Annals of Tourism Research* 19:1, 68-78.

Olson, M. and Crase, D. (1990), "Presleymania: The Elvis Factor," *Death Studies* 14:3, 277-82.

Reader, I. and Walter, T. (eds.) (1994), *Pilgrimage in Popular Culture* (London: Macmillan Press).

Rodman, G.B. (1994), "A Hero to Most? Elvis, Myth, and the Politics of Race," *Cultural Studies* 8:3, 457-83.

Rodman, G.B. (1996), *Elvis after Elvis: The Posthumous Career of a Living Legend* (New York: Routledge).

Seaton, A.V. (1999), "War and Thanatourism: Waterloo 1815-1914," *Annals of Tourism Research* 26:1, 130-58.

Schein, R.H. (1997), "The Place of Landscape: A Conceptual Framework for an American Scene," *Annals of the Association of American Geographers* 87:4, 660-80.

Wright, D. (1996), *Dear Elvis: Graffiti from Graceland* (Memphis, TN: Mustang Publishing Company).

Chapter 5

Ambient Australia: Music, Meditation, and Tourist Places

John Connell and Chris Gibson

Introduction

This chapter examines how music informs the creation of tourist places in Australia. It discusses one genre—ambient music—and the way it is related to geography both symbolically (in terms of cultural representations), and literally (in terms of links to musical and touristic activities in particular towns). The rise of ambient music has contributed to the imaginative representation of a touristic Australia of "natural" physical and cultural landscapes, where indigenous people are particularly significant. Designed to encourage relaxation and even sleep, in its cover art, its sounds and lyrics (where they exist), ambient music has emphasized "special" places both generic and real, that are remote from urban centers, and physically attractive—usually involving mountains, falling or flowing water (in streams rather than rivers), rain forests, coasts, seashores and oceans, occasionally deserts, and more generally "wilderness." Ambient music is thus a means through which a very particular cultural geography of landscapes and nature is constructed and vicariously experienced. Landscapes are imbued with certain spiritual powers, or associated with animals, such as birds, dolphins and whales, regarded as having special qualities, the sounds of which are incorporated into many tracks.

Ambient music is also now a distinct market for particular musicians and record labels, which have emerged to satisfy tourist demand for aural souvenirs of Australia. Companies not normally associated with music have established distribution networks that center on popular holiday towns, bookshops, and airports rather than music stores—and that rely much less on digital download technologies than other mainstream music genres (though some ambient artists such as Ken Davis have made tracks available to download via iTunes). Indeed, because ambient music is a souvenir as much as anything else, the material object of the CD matters a great deal, and infiltration of downloading as a regular music consumption practice is far less widespread than in mainstream markets. Buying a track or two online after a holiday is simply not the same as purchasing the CD in situ as a souvenir.

In part because of this, in certain material geographical locations the presence of tourism industries *and* local, vernacular cultures has fuelled specific relationships between ambient music and place. Byron Bay, once a small coastal town centered on a whaling station, is now a site of global backpacker culture, and has a growing music scene connected to it. Ambient, "world" music, and Aboriginal sounds are featured in local retail outlets, workshops (from tantric drumming classes to didjeridu lessons) and performances, providing confirmation of the "new age" symbolism of ambient music recordings. Conversely, Alice Springs, in Central Australia has become a site of Aboriginal cultural tourism, where didjeridus and "Aboriginalized" ambient music are popular. In Alice Springs there are possibilities for Aboriginal musicians to participate in the tourism industry because of demand for music, yet these opportunities are circumscribed by difficulties in competing with metropolitan-based souvenir companies who produce and distribute CDs of "Aboriginalized" ambient music, or with non-indigenous musicians (who have more resources and marketing skills at their disposal). In these ways, ambient music—a unique form of music production, marketing and consumption—combines with tourism to transform places discursively and materially.

Despite growing interest in the geography of music (e.g. Kong 1995; Smith 1997; Carney 1998; Connell and Gibson 2003), there are still types of music under-explored, notably ambient music. Ambient music is often derided and ignored in comparison to other genres because of perceptions that it is trivial, authorless, repetitive, and mere mindless, background music. Yet, ambient music has a distinct history (Lanza 1994), and like other music styles, it is entangled in a geography of place, landscape, and identity (Connell and Gibson 2003). In Australia, ambient music also exists in a particular structure of tourist marketing. Music plays a part in creating tourist destinations (Atkinson 1997; Gibson and Connell 2005; Cohen 2007), and music itself is often made with tourist audiences in mind (Kneafsey 2002). A continent long-imagined as remote, wild and untamed, tourist Australia is the antithesis of northern industrial modernity. Ambient music, marketed particularly to foreign tourists, plays on this imagination. Ambient music both creates images of places, contributes to the selling of those places as tourist destinations, and constitutes a new form of physical musical commodity that in certain circumstances is linked to the marketing of actual musical performances in destinations, and thus is linked to the transformation of touristic urban spaces.

This chapter discusses ambient music, and its relationship to an imagined Australia that features in the touristic construction of nationhood, and in new kinds of retail urban spaces geared towards tourists. The first section of the chapter outlines the emergence of ambient music as a genre, and is followed by interpretations of the links between ambient music, place, and tourism. To highlight the links between ambient music and tourist places, we examine contemporary transformations of the two tourist towns mentioned above: Byron Bay, on the New South Wales far north coast and Alice Springs, in central Australia. In doing so, we seek to combine representational and material approaches, blending cultural interpretation (framed around representational analysis of place identities) and political economic analysis (through the tourist industry and retail market for ambient music in Australia). Running throughout is an argument about cultural essentialism and the social construction of nature (see also Connell and Gibson

2008), both of which are apparent in the representational elements of ambient music, and in the orientation and intentions of marketing "real" places to tourists in Australia.

Origins of ambient music

Ambient music coalesced as a discernible music genre in the 1970s in the West, with the rise of new age culture and the emergence of specialized labels releasing recordings designed for relaxation and background music. For many, the apex of ambient music arrived with the release in 1978 of Brian Eno's seminal *Music for Airports*, which bestowed credibility and critical purchase on what had previously been considered a tacky musical genre, and exposed a different audience to new age releases. The invention of Muzak in the 1930s ushered in the distribution of music primarily for playing in the background to domestic situations, rather than as an object of "active" consumption. Likewise, through the 1950s and 1960s countless "easy-listening" records were released by artists such as Martin Denny, Bert Kaempfert, and the Melachrino Strings, designed to create moods in the home rather than to be listened to attentively. It was only in the 1970s that "ambient music" emerged as a term to describe a particular musical form, designed to encourage relaxation and meditation, whilst also being invested by its practitioners with particular healing qualities. Whereas easy-listening records were generally light cover versions, or caricatured ethnic difference (Lanza 1994), ambient music was often linked to new age philosophies, renewed interest in traditional cultures, esoteric book stores, herbal remedies, crystals, and a range of unorthodox spiritual beliefs. At the same time it constituted a new kind of background music, which again was usually seen to have certain spiritual properties.

Ambient music has become particularly popular in Australia, sold on CD format almost exclusively, and with releases usually undated, to maintain their longevity as souvenir objects (being a "latest release" matters far less in this market; tourists buy an ambient CD not because it is brand new, but because of its association with a place or touristic experience). CDs are marketed in tourist destinations (including airports) and stimulate a distinctive touristic geography of Australia. The rise of ambient music has contributed to the creation of an essential "natural" Australia of mountains, seas, and rainforests where indigenous people are of particular significance. Though rarely to be heard on the radio or in live performance, and derided or totally ignored by most music critics, it has achieved spectacular sales with performers such as Ken Davis, Tony O'Connor, and David Hudson achieving multi-platinum status with several recordings. Some Australian artists have been particularly successful overseas, further suggesting the link between music, memory, and place. Steve Deal, whose music sells particularly well in Germany, has recorded a series of albums including *Spirit of Australia*, *Didgeridoo Dreaming*, *Native Birds* and *Outback*, but is largely unknown in Australia. Using marketing strategies rather different from those operating in the wider music industry, specialist labels and souvenir companies have come to dominate ambient music in Australia—shaping what music is made available, especially in tourist destinations, and with important implications for indigenous musicians.

"Natural" Australia: An ambient sense of place

Australian ambient music emphasizes "special" places, both mythic and real, that are iconic, remote, physically attractive, and seen as possessing spiritual powers. As with ambient music elsewhere, "natural" sounds are incorporated into many tracks. Thus, the liner notes to Andy Holm's *Blue Mountain Tales* (1996), on Primal Harmony Records, stress the significance of the Blue Mountains, immediately to the west of Sydney:

> During my frequent travels to the Blue Mountains I have always been inspired to pick up my instruments and play out the feelings that arise within me. This album is the product of the precious moments. Whether I played my flute at the bottom of Wentworth Falls or the didgeridoo at Echo Point, the vibration of the surroundings always influenced a strong tribal and spiritual sound, a reflection of my inner self opening up at the Blue Mountains beauty. Although this album is a fairytale story, it is reflecting on my true emotions brought forward by the ambience of the Blue Mountains. Close your eyes while you are listening, let the narrator be your guide and join me on a magnificent voyage through time, sound and space, and a dimension you will find within yourself listening to this work.

Similarly, other prominent Australian natural icons have been the focus of recordings. Some performers actually recorded within the natural landscape, but more frequently entitled their albums with the names of specific places, such as Uluṟu (Ayers Rock) and Kakadu (the name of a world heritage national park in the Northern Territory, known for its Aboriginal cultural heritage), that had particular natural and spiritual significance. A link to these places—which are also major tourist drawcards—is one means to enhance the possibility of success.

Although specific places are celebrated regularly in this way in Australian ambient music, they are few in number. In his *Celebration* series Tony O'Connor includes *Windjana—Spirit of the Kimberley*, about a remote region on the northwest coast of Australia. The album is described in the liner notes as "a tribute to one of the Earth's last untouched wilderness areas" (which it is certainly not, because it has a long history of Aboriginal occupation, and subsequently has been a site of European pastoralism and mining; cf. Cronon 1996). On *Uluru* (referring to the iconic megalithic rock formation in central Australia), didjeridu sounds are combined with "soaring orchestras" and pan flutes: "*Uluru* has a movie soundtrack feel to it, yet remains soothing and relaxing ... I always imagine eagles soaring or wide aerial shots over the vast red sands of central Australia," according to the liner notes. Ken Davis's *Spirit of the Outback* (2006) incorporates such tracks as "Nightfall over Uluru," "Dreamtime Spirit" (a reference to a specific Aboriginal spiritual tale about how land was created), and "Changing Colours of the Rock" (again, referencing Uluru). O'Connor has also recorded *Kakadu*. The Great Barrier Reef is celebrated in Tania Rose's *Coral Sea Dreaming* (1996) which is also linked to a nature documentary about the Reef. The CD cover argues that "until we learn to see wonderingly and non-possessively, it is not only the future of the world's natural environment that are threatened but our own survival."

Alternatively, nationhood (and souvenir quality) can be linked in ambient music to more generic natural environments. Australian musician Tony O'Connor, whose website proclaims that he has sold three million recordings, has produced a six part *Celebration of Australia* series, including his best-selling *Rainforest Magic*, which through "a distant waterfall, a gently babbling stream … [provides] … an adventure in nature's wonderland." The album series followed a year's travelling in Australia, including such rainforests as Daintree, Lamington, and Tasmania. This was then followed by studio time to "try and recreate that special stillness and magic of the forest." Other recordings in the series similarly emphasize generic natural landscapes, including *Whispering Sea*, *Wilderness* and *Hidden Forest*; the notes for the last state that "In a hidden valley near a cave at the bottom of a waterfall, there is a special secret place—a place you can go and relax and hear the whispers of a forest dreaming." Ken Davis's *Australian Atmospheres* (2004), which includes such tracks as "Waratah and Wattle," "Desert Moon," "Coral Reef Dream," and "Pacific Sunset," suggests in the liner notes to "let these timeless instrumentals inspire you of the magic of *The Great Southern Land*." For the foreign tourist, generic references to Australian nature are sufficient to evoke memories of their holiday—background music as pleasant souvenir.

That places are and should remain pristine is evident in the choice of landscapes and sounds evoked—whether specific or generic. Ken Davis goes beyond this on his website where different pages conclude with evocations to "Plant Millions of Trees so that we can all Breathe" and "Protect Whales and Dolphins and all Endangered Species." Ambient music proffers a soft variation of green philosophy—Gaia theory without the radical politics—that emphasizes the need to preserve Australian wilderness, to connect with other living beings, and to appreciate the unchanging, timeless character of nature. Other realities of Australian "nature"—salinity, old growth logging, bleaching of coral from pollution, soil erosion in marginal landscapes—are not surprisingly absent, replaced instead by an idyllic nature free from human influence. Robert Rankin's *Wilderness* (1994) exemplifies this, accompanied by a high-quality booklet of ten professional photographs (by the composer, whose main endeavour is photography, but who also sells calendars, diaries, books, and posters) of the landscapes inspiring its six tracks ("To the Far Horizon," "Ocean Sunrise," "Sandblow," "Lamington Rainforests," "Purnululu," and "The Kimberley"). Human figures are nowhere to be seen.

The spirit of place

If Australian nature is "wild" and humans are absent from the landscapes depicted in much ambient music, ambient music also somewhat contradictorily positions the human in relation to the specific nature imagined in its marketing, cover art, and sounds. Sometimes this is overtly the case—as we discuss below in relation to the depictions of indigenous peoples—while in other examples, the link is more subtle. Ambient music stresses the mystical and "cosmic" life forces that are associated with "traditional peoples," their links to the earth and their musical instruments. In Australia the didjeridu (often misspelt "didgeridoo") features in much ambient music, especially that of the Aboriginal performer David Hudson, but panpipes, and also flutes (seen as primordial musical instruments evoking pre-modern humanity), are common (though not necessarily tied to their geographical origins). Ken Davis, who is not Aboriginal, in his *Spirit of the Pan Flute Australia* (2003; a CD featuring Uluṟu on the cover) proclaims "Haunting pan flute melodies and didgeridoos create the Spirit that is unique to our country. Visit the many places through music of this incredible place." Beyond generic indigenous peoples in global ambient music, other humans who have particular spiritual powers might also be featured, including shamans on some North American music. Natural landscapes and "laid-back" music, incorporating indigenous instruments and sounds (invoking an explicit conjunction of indigenous people as "natural") have contributed to primitivist fantasies of tranquility, timelessness, and innocent human interactions with nature. The point is that listening to this music is meant to bring relaxation, spiritual healing and escape from modern, industrial, urban life—vicarious tourism to an ancient Australia free of contemporary socio-economic and environmental problems, in which humans tread lightly across a pristine nature. This is precisely the formula through which Australian governments have sought to market the country to foreign tourists in television and print advertising campaigns (Waitt 1997; 1999).

Accordingly, the cover of *A Voyage to Australia* (1998), a CD that like many others provides no indication of the performers, authorship, or origins of the music, proclaims:

> Australia; a country of green hills and fertile soil. A country of dense rainforests and streaming monsoon rains. But also a country of barren dryness and endless plains. An island where extremes come together. It is the country of numerous aboriginal tribes, the original inhabitants of the continent. For tens of thousands of years they have lived here in complete harmony with nature. They know that mankind does not own the earth but that the earth owns mankind. In the dreamtime the gods created the land the rocks, the rivers, plants, animals and mankind. The aboriginals [sic] still describe the creation in their music and in the tales that are passed from generation to generation.

Tracks on the album, clearly performed by un-acknowledged indigenous musicians (there is no sleeve information provided on who the performers are, but from the languages sung and instruments used, they are clearly Aboriginal performers with strong tribal backgrounds), include "Nangalor Dreaming," "Gunbalanya," Kalkarindji's Walkabout," "Corroboree Rhythm," "Footsteps in Red Soil," and "Mystical Uluru." They are all intended to be indicative of connections with Aboriginal pasts, emphasizing timelessness and the link between people and nature unmediated by modernity, and especially evident in the cover design. (The spelling of the final track, "Maze of Wheathered (sic) Sandstone Domes," is indicative of the often low production standards of CDs designed for an instant and uncritical tourist market). Such imagery is in danger of rendering indigenous people "exotic" but silent objects of neo-Darwinian landscapes (Neuenfeldt 1994)—humans further back on an evolutionary scale, closer to nature, and objects for the tourist gaze and the vicarious tourist ear.

While non-indigenous performers have increasingly become cautious about stressing any indigenous heritage within their music, Aboriginal performers have not surprisingly been less reticent. David Hudson, from the western Yalangi people of north-eastern Queensland, argues on his website that his music is "contemporary" but with "traditional elements" and is "inspired by the simplistic [sic] things like the running water of a mountain stream, leaves falling in the rainforest, and the rich cultural history of his ancestors—the indigenous people of Australia." However his music goes far beyond the land of his own cultural group. Proclaiming himself on his website as "the ancient voice of the future," Hudson "takes you on many journeys throughout this vast country, from rainforest, to the reef, to the outback." Hudson's music is primarily based on the didjeridu, sometimes accompanied with other instruments. It ranges from albums such as *Kuranda* (1997; named after the place of the same name near Cairns in far north Queensland—both heavily frequented by international tourists) and *Guardians of the Reef* (1996; again a specific place reference to the Great Barrier Reef) to a version of Puccini's operatic aria *Nessun Dorma* (1997) played with saxophone, guitar, and didjeridu. Another Aboriginal musician, Ash Dargan, from the Larrakia cultural group in the Northern Territory, claims on his website to have "woven a tapestry of music that echoes with ancient voices and has the power to send audiences to Dreamtime states" in a genre described as "world indigenous big beat."

Both David Hudson and Ash Dargan release recordings through the Indigenous Australia label, originating in Cairns, but now based in inner-city Sydney, that has also released albums such as *Flamenco con Chilli: A Hot Flamenco Blend of Modern Flamenco and Ancient Didg, Island Warriors (music by Kara Kazil, "traditional people" from the Torres Strait Islands)* and *Rio Didg: A Latino Excursion*. Their website (www.indig.com) greets visitors with a recorded voice saying "Welcome to indig.com, home of the world's number one independent indigenous world music experience. Indulge yourself in the sounds of indigenous cultures from around the world." Another stream in its releases focuses on neighboring New Zealand ("take a journey into a timeless soundscape of inspirational voices and sounds, reflecting the pristine beauty of New Zealand"). In this instance, ambient music fuses with the marketing of "world music" more widely (Connell and Gibson 2004). Fragments of "traditional," "indigenous," or "ethnic" music are assembled and melded in differing combinations for primarily western, touristic consumption. The human is everywhere in this version of ambient music—but selectively sampled and reproduced as timeless, pre-modern culture for jaded urban consumers.

Similar timelessness is evident in David Hudson's *Guardians of the Reef* (1996), a CD that combines flute and didjeridu:

> A long time ago, when the land was one, to enter the region that is now the Great Barrier Reef, one had to be granted permission by the Guardian Spirits. Once there one had to live in harmony with the earth and learn to exist joyfully, side by side with his fellow creatures. The Guardians of the Reef kept a watchful eye on the creatures to keep them safe and free from harm so the reef would be there for all to enjoy for eternity (CD liner notes).

While some statements of this kind may have some basis in Aboriginal legend, contemporary ethnographic accuracy is barely present.

Beyond relaxation, some ambient music expressly claims healing qualities, or at least claims to be an adjunct to healing practices that are encouraged as borrowings from indigenous and non-Western cultures. Tony O'Connor's *Awakenings* has seven tracks, each of which are seven minutes long, and "each one dedicated to each of the Chakra energy points [vortices believed in Indian medicine to be present at various points across the human body]." With a didjeridu as the key sound it is claimed in the liner notes to be "a wonderful album for masseurs or for anytime that you wish to enter a meditative state." O'Connor, who lives on an organic farm in southern Queensland, has expressly claimed that the structure of his music was derived in part from working with someone writing a thesis on the physiological effects of music. Somewhat similarly, promotional literature for John Keech's *Pure Meditation* (2003), written by an expert identified as having a BA, MSc (in Psychology) and PhD, claims that it is a

> welcome contribution to the field of quality relaxation music. Its special features allow the music to support a wide range of relaxation imagery, making it suitable for both newcomers to the practice of music assisted relaxation, and to those who already have a favourite relaxation routine.

The website of Tony O'Connor includes a guestbook where listeners can record their thoughts on his music. Thus, Clare Power from the Isle of Man registers:

> Australian friends introduced me to your music some years ago ... I introduced the wonderful flowing sounds to the T'ai Chi Chuan Group, for Shibashi practice. Also your *In Touch, Bushland Dreaming, Kakadu* and *Uluru* keep me in touch

Figure 5.1   Indig.com website

Source: Reprinted with kind permission of Indigenous Australia, Pty Ltd

with my daughter's family and my friends "down under" … Thank you again, Namaste.

Emigrant Australians, such as Robert, an "Aussie living in Denmark," may "listen to your music when I am homesick." An American, in Texas, writes: "I have found it very inspirational and beautiful, just like your homeland." Most writers simply focus on the music and its soothing values: Kristin writes from the United States that "There are so many stresses in this world and all too often even the music we are bombarded with is equally stressing. It is so nice to have an artist such as yourself creating such a safe haven to go to." Music provides the promise of emotional transformation linked to evocations of nature (fauna, flora and landscapes), timeless cultures, and ancient, iconic, and idyllic places.

Transformations of touristic space

Ambient music stresses virtual tourism, both in its content and the marketing, and unlike other genres it is sold mostly in bookstores and souvenir shops in tourist venues and airports. In Alice Springs ambient music is sold alongside Aboriginal artefacts (Dunbar-Hall and Gibson 2004). Much is sold from interactive displays where samples of the music can be heard pre-purchase (Figure 5.1). Such interactive displays constantly play to passers-by, broadcasting encouragement to stop, listen to, and buy the music. The first such interactive unit was in the Australian Geographic retail outlet in Darling Harbour, Sydney, in the early 1990s. The principal ambient distribution company, Holborne Australasia, now has more than 1,500 units in a series of "non-traditional" retail outlets in Australia including gift and tourist stores, news agencies, bookstores, and airports. Virtually no ambient music of this type is sold in conventional music stores.

Indeed, the system of production, distribution, and marketing for ambient music is much less formalized than for conventional music, and it is connected to sets of allied industries (in tourism, souvenir production, and publishing). Whereas conventional music is still distributed through a heavily oligopolized cartel of major labels (Sony-BMG, Universal, Warners-EMI) to established retail chains (HMV, Sanity, JB hi-fi), and supported by an accompanying marketing apparatus (commercial radio, television music shows, cable music video), the rise of, and subsequent touristification of ambient music has been shaped by more diffuse players operating in a range of non-conventional spaces and networks. Digital downloading has not transformed the ambient music industry to anywhere near the same extent as for other genres. In a parallel to the conventional music industry, one company—Holborne (that also controls and distributes the Indigenous Australia label, home to David Hudson and Ash Dargan)—dominates the distribution of ambient music, through its in-store marketing and interactive listening stations. According to Indig.com: "In less than two years David Hudson's *Touching the Sounds of Australia* series has collectively sold over one million units globally, making him one of the most successful Australian recording entities of all time. The company's newest series, *Indigenous World Rhythms of Australia* … has already topped 100,000 sales."

A small number of individual recording artists have been able to remain strong players in the market for ambient music, amidst this monopolization. Tony O'Connor and Ken Davis have released recordings themselves, often selling copies in hundreds of thousands (multi-platinum in Australia), in the case of the former through a souvenir company in his name that also distributes calendars, cards, and posters. David Hudson's web page firmly links the music and tourism, noting that he will also take orders for paintings and didjeridus, which "include an authenticity certificate signed by David." Other links from the site sell such artifacts as mouse pads, pewter, leather, opal jewelry, and boomerangs.

The links to tourism are not only manifest in the relationships between ambient music distribution companies and the tourism industry, but in how certain destinations have themselves been reshaped in light of the images of place represented in ambient music. Music is important for tourism destinations in a number of ways. Where places are known for musical heritages, branding strategies capitalize on this to attract potential visitors (Cohen 2007; Atkinson, 1997; Gibson and Connell 2007; Connell and Gibson 2008). As at Memphis, New Orleans, Nashville, Hawai'i and Liverpool, past or current musical production, made famous internationally, defines regional identity.

Tourist attractions have been created in locations to cater for visitor's demand for music reflective of popularized representations of place: in Nashville ("home" of country music) and Pigeon Forge, Tennessee (the birthplace of Dolly Parton), a range of country music theatres, museums, a hall of fame, and even theme parks have been built; while in an entirely different context, classical music tours have revitalized festivals and music halls in cities such as Vienna, Prague, and Salzburg (Connell and Gibson 2003; Gibson and Connell 2005). Music also plays an important role in tourism economies as a component of visitor activities, whether officially-promoted within wider strategies (as with festivals and special events), or as part of regular tourist schedules within particular market segments (as with young travelers and dance music in locations such as Ibiza, Spain and Goa, India). As well as playing an important role in mainstream tourism promotions, music can also form the basis of informal economies (buskers, local gigs, street performers). While having little direct impact on marketing strategies, such activities can provide local musicians with a "top-up" income, and significantly enhance local cultural identity and marketability for regions, providing a sense of "existential authenticity" (Wang 1999) in the landscape—embodied, spontaneous cultural interactions for visitors, beyond more formal promotions, events, and tours. Music, previously underestimated as an element of cultural tourism, has formed a means through which landscapes have been promoted and transformed.

The two examples discussed briefly here—Byron Bay, on the New South Wales subtropical far north coast, and Alice Springs, in central Australia—are popular tourist destinations where a symbiotic, dialectic relationship exists between ambient music and the tourist industry. They are places that have been reshaped in light of images of place popular in ambient music, and because of associations with new age and Aboriginal culture respectively, have become places where ambient music is sought, and actively marketed.

Byron Bay

Byron Bay, a small ex-whaling and abattoir town (and now host to thousands of international backpackers) has become a crucial site of interaction between tourists and music scenes (notably in techno, folk, blues, and "world" music; see Gibson and Connell [2003] for an extended discussion of this case study). In Byron Bay, music has emerged as a by-product of the growth of "alternative" subcultures since the 1970s (as it became the site of countercultural migration), shaping images of place that have made the town and region more marketable as a tourist destination. In turn, the backpacker tourist market has formed a crucial stimulus for new venues, festivals, and performances.

Nearby Nimbin became a haven for alternative cultures in the 1970s, when the National Union of Students held its inaugural Aquarius Festival (Australia's answer to Woodstock) there in 1973. This triggered a wave of inter-regional migration to the area from capital cities, as others sought to "drop out" by moving to the region, finding one of the area's communes to live on. Byron Bay became

Figure 5.2   New age retail outlets, Byron Bay, NSW, July 2004

known as a place where surfing and new age cultures mixed. It was incorporated into the backpacker circuit (that grew substantially in the 1980s), and became a place of rock festivals, folk performances, illicit drug consumption, new age culture, shopping, and beach lifestyles.

Byron Bay was thus a ready market for ambient music. In the 1980s shops selling crystals, taro card readings, "tribal" artifacts and fashion began distributing ambient music, didjeridus, and drums (see Figure 5.2). Customers were both locals participating in new age culture, and tourists attracted there because of the same associations.

A number of musicians from the area began to compose and independently release ambient music, including Dr Didge, Si, Jalalo, Coolangubra, Ganga Giri, and Tarshito. Their releases were targeted at tourists, sold in new age bookshops, souvenir shops and at local craft markets—frequented by backpackers in particular. Common features of ambient music in Byron Bay include folk music influences, generic environmentalism, the use of a wide variety of instrumentation, particularly percussive instruments from different national/ethnic contexts, widespread use of synthesizers, and also at times a strong presence of a capella vocal arrangements (for example, the Voices of Gaia). As with more successful national artists, the emphasis is on relaxation, spiritual awakening, and vicarious travel to "other" places and states of mind. Again, a link with nature is vital, as with Jalalo's *Letting*

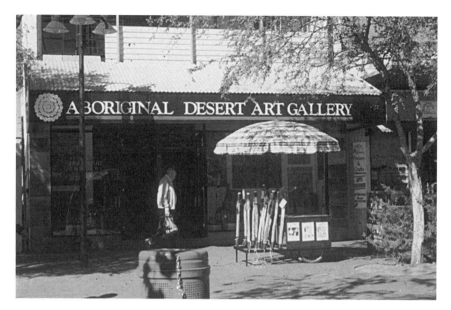

Figure 5.3   Todd Street Mall, Alice Springs, July 2001

Go (1995): "Nature inspires a lot of my work … very often I sit with my keyboard looking out the window. That puts me in the space for writing. Often I get ideas for music, walking in nature. I hear it all in my head. Nature is a very strong inspiration for me" (*Our Times* 1996, 11). Similarly, the Windsong Records web-label, specializing in new age releases, advertises through connections to place. According to its website, "Byron Bay is a cultural and creative melting pot; a favourite destination for many a world traveller. Many artists have made this region their home, a place where they can be inspired by their surroundings and pursue their passions." In this niche, depictions of local beaches, ocean scenes, rainbows and dolphins are common, images that also dominate official tourist promotions for the region. These abound in ambient releases by David Birch (*Byron Journey* 1997), and the prolific Tarshito, who has to date released over 15 full length CDs and cassettes. Ambient music's predilection for idyllic natural landscapes, new age mysticism and tribal culture coalesced to form the basis of a distinctive local creative industry. This in turn helped transform a small prosaic coastal town into a new age tourism destination.

Alice Springs

At Alice Springs (the second largest town in the Northern Territory), in the central Australian desert, ambient music, alongside other creative arts has also played a role in creating images of place, in promoting particular types of tourism activities, and in transforming its urban landscape. Yet in Alice Springs, its remoteness and status as a contemporary center and meeting point for Aboriginal communities also shapes the local contexts of cultural tourism. Like "traditional" tribal visual art, music is a growing cultural industry for Australian Aboriginal communities, and as such is bound up in a range of issues including the ownership and use of cultural resources, representations of place, the distribution of benefits, and extent of Aboriginal employment (Gibson and Connell 2004). Moreover, Aboriginal cultural expressions such as art and music have been translated into the materiality of the town's streetscapes (Figure 5.3), reflecting what Anderson and Jacobs (1997, 19) have called the "Aboriginalization of urban space," emanating from non-Aboriginal intentions to "produce developments and planning initiatives which actively seek to include some type of Aboriginalized aesthetic or content." Alice Springs has grown an "Aboriginalized" tourist economy and built an "Aboriginalized" urban landscape through art galleries, souvenir shops, public art, and ambient music.

While natural landscapes have been central to marketing inland Australia in tourism campaigns, cultural artifacts have been strongly associated with imaginings of Aboriginal society, particularly art, dance, crafts, and music. Nearly three-quarters of international visitors cite Aboriginal art and culture as the activity of highest interest; while more specifically, 56 percent of all international visitors purchase Aboriginal cultural items such as art, crafts, and music (ACNielson 2000), worth over A\$50 million per annum (Northern Territory Government 1999). The links between Aboriginal expressions, tourist-oriented landscapes, and cultural industries is governed by relationships between producers, distributors, marketing bodies, and tourists—the final consumers. These relationships are simultaneously cultural and economic: Aboriginal authors, musicians, and craftspersons create works for a number of cultural and commercial reasons, sometimes as part of localized traditions, or as a part of "scenes" (such as Aboriginal bands and performers; see Gibson 1998), but also in response to perceptions of tourist demand for cultural products. Commodities such as art and crafts may be created in such a way as to be strategically (in)authentic—appealing to generalized "traditional" styles and designs in response to the needs of the tourist market, but also as a means of generating wider knowledge and understanding of Aboriginal cultures. Conversely, appropriations of indigenous designs and styles by non-Aboriginal producers have been widely criticized as acts of cultural theft, most famously in the mass production of consumer goods such as t-shirts, posters, and carpets. As has been documented elsewhere (Janke 1998; Johnson 1996) the moral rights of Aboriginal authors have been systematically denied in Australia, and communities from which designs originated have been excluded from the benefits of commercial sales. In short, there are a range of "benefits and disbenefits associated with specific kinds of commodification" (Jackson 1999, 97), nowhere more apparent than in the links between Aboriginal cultural production and tourism.

Music has formed a crucial part of this nexus between Aboriginal culture and tourism. Aboriginal sounds—most familiar in traditional singing styles and in certain forms of instrumentation such as the didjeridu and bilma (clapsticks)— have become a soundtrack to tourism marketing (in television advertisements, for instance), as well as occupying a segment of the market for actual artifacts sold to tourists. Certain "Aboriginal" sounds, usually traditional didjeridu and vocal pieces originating from the Top End (the northernmost parts of the Northern Territory and far north Queensland) have become aural signifiers of Aboriginality in ambient music, and the associated tropes of a static, ancient culture. Hence *The Very Best of Ash Dargan* (2007) is described as "A rhythmical walkabout with the pulse of life through the Top End's pristine Dreaming Country. Inspired tracks from the spiritual home of Ash Dargan – the Top End, Australia's Northern Territory."

In particular, the didjeridu has become popular among backpackers from Europe and North America. The didjeridu has become an artifact imbued with global cultural resonances, as it has surfaced in "new age" discourses of healing and spirituality (Neuenfeldt 1998), and become a worldwide symbol of a romanticized indigenous past, "a musical instrument of limited historical distribution has been globalized to become, for some users and audiences, a metonym for Aboriginal culture or indigeneity as an abstract whole" (Neuenfeldt 1997, 108). Hence a certain confluence of two otherwise distinct narratives occurs: music contributes to representations of Aboriginality in tourism promotions, yet at the same time those Aboriginalities reinforce particular semiotic constructs in music itself.

The ubiquitous practice within the ambient music scene, where music is argued to have some link with Aboriginal cultures, of ignoring authors and the origins of the music, and in some case the performers, denies the authority of Aboriginal composers in the production of music, amounting to cultural theft. Thus *Woomera* (1999), produced on the Indigenous Australia label, is simply subtitled "Traditional Aboriginal Music" but, other than noting Ash Dargan's involvement on "didj," "composition and music" are credited to Nigel Pegrum and no individual Aborigines or cultural groups are mentioned, hence none have access to royalties (despite the obvious traditional names of songs on the recording).

Moreover while such music is marketed as being the authentic sounds of the indigenous peoples of a timeless Australia, notably in the main tourist mall of Alice Springs and in the Holborne retail listening outlets around Australia, the bulk of the contemporary music that actually is produced by Aboriginal Australians is ignored and excluded. Thus, for more than twenty years CAAMA (the Central Australian Aboriginal Media Association), also based in Alice Springs, has produced and marketed Aboriginal performers in a range of genres, from country to heavy metal, but a very large part of these recordings have failed to make the short journey from their recording studios in Alice Springs to souvenir shops and Aboriginal art and craft centers in the town's tourist precinct, so that CAAMA themselves estimate that Holborne control about 90 percent of the tourist market for Aboriginal music in the town (Dunbar-Hall and Gibson 2004). The proportion is very much higher, and probably 100 percent, in most other tourist destinations. In two parallel ways, and despite the existence of David Hudson and Ash Dargan, the rise of ambient music has excluded, marginalized, and denied contemporary Aboriginal people and their music, all the while distorting and glamorizing "traditional" Aborigines and their sounds for the tourist gaze/ear.

Ironically, at Uluṟu, the quintessential tourist site that is visited from Alice Springs, many tourists blithely deny Aboriginal requests that they not climb the rock for spiritual reasons, emphasizing the wider conflict between commerce and culture in the region (Digance 2003; Waitt et al. 2007). Cultural tourism experiences in this iconic Aboriginal landscape—including visiting sacred sites such as Uluṟu as well as listening to music and watching musical performances—may be a form of cultural learning, even a pathway to reconciling western and indigenous cultural values. However, as is patently the case with the marketing of ambient music to tourists in central Australia, monopolization of distribution channels and discourses of what "indigenous music" is meant to sound like to the tourist ear, circumscribe the potential for resulting economic benefits to be shared more equitably amongst local Aboriginal musicians.

Thus, a very particular fragment of Aboriginal music (didjeridu-based) has become a cultural resource available to the tourism industry both as a promotional text (signifying the ancient and tribal), and part of a distinct set of contemporary products, experiences (performances) and objects (recordings, didjeridus as souvenirs). Moreover, music has distinct spatial qualities that allow it to transform material landscapes—filling urban spaces with sound, most often by buskers, imbuing those spaces with particular meanings as tourist-friendly zones. In Alice Springs the visual and aural are combined to construct an Aboriginalized tourist business district. Todd Street, the town's main street, has become a pedestrian mall, shut off to vehicular traffic in the 1980s, when it was originally designed in an Anglo-Celtic "colonial frontier" theme, with iconic windmill and native trees, shop fronts with extended "Georgian" style verandas, and widespread use of rustic furniture, border edging, and signage. Todd Street Mall over time attracted a mixture of retail outlets increasingly geared towards tourists—as well as functional businesses (banks, chemists, etc)—and it now features cafes, restaurants, pubs, art galleries, art shops, souvenir shops, travel agents, safari gear, and camping shops. Alongside ubiquitous souvenirs is a range of "Aboriginalia" (Neuenfeldt 1997), from didjeridus to drums, paintings, pottery, t-shirts, posters, and necklaces.

Ambient music is an important part of Aboriginalizing this tourist landscape: art shops play ambient recordings of didjeridu as an accompaniment to selling art—providing a solemn backdrop and seemingly affirming authenticity of objects for sale. Meanwhile buskers (of both Aboriginal and non-Aboriginal background) play didjeridus to backing tapes of ambient sounds; souvenir shops place Holborne interactive CD displays prominently at the front of stores, and more "serious" performances are staged in specialized venues built along the street specifically to cater for tourist demand. An example of this is non-Aboriginal local performer Andrew Langford's Sounds of Starlight Theatre. In his own words on a flyer touting the productions to international tourists, the productions staged there present

> a musical journey through the ancient landforms, history, light and space of Australia's heartland ... Andrew Langford demonstrates his mastery of the didgeridoo to evoke a vibrant wilderness of sandy deserts, mountain ranges,

deep gorges and icy waterholes. He captures the contrasting power of Ancestral Beings and the rhythmic march of honey ants.

Continuing the spiritual theme, "Andrew is supported by guest artists using an array of world instruments including the Aztec hum drum, spirit catcher and shamanic drums used by American Indians and the qweeka of Latin America." The characteristics of the wider ambient music market—an emphasis on nature, wilderness, primitivism, pre-modern spirituality—infuse this actual tourist attraction in a place iconically associated with Australian tourism.

While Aboriginality is now thoroughly commoditized and seen (and heard in the case of ambient music) in the retail landscape of Alice Springs, the regular presence of Aboriginal people in the same Todd Street Mall has caused tensions, particularly with shopkeepers who resent public drinking by Aboriginal people, and resulting confrontations with tourists. Police now more stringently survey the mall, breaking up groups of Aboriginal people and removing those considered to have committed crimes or engaged in "anti-social behavior" (such as begging, fighting, and drinking). All the while, the ambient sounds of didjeridu and clapstick, and also of synthesizer, pan pipes, and flute, are broadcast from the mall's background sound system and from shops selling Aboriginal art, music, and souvenirs. In its role in tourism promotion, ambient music has overwhelmed, distorted, and even denied the contemporary Aboriginal presence in Alice Springs—a strange soundtrack to a place made by, and yet continually evicted of, its indigenous cultural significance.

Conclusion

> In the 1990s, new age music mutated into more generic relaxation. This insidious marketing ploy was particularly prevalent in Australia. Fans of quality ambient were no doubt aghast as, via post offices and bookshops, the market was flooded with extremely suspect nature-themed recordings. These recordings sold very well and made some very mediocre musicians lots of money. Tourists love it, too, with the slickly packaged images of Australia's natural heritage proving very attractive. But far from Eno's vision of ambient, this music was not 'simultaneously relaxing and engaging'. Most of it was just plain awful. Perhaps knowing this fact all along, many of the perpetrators simply took the money and ran as the music's popularity inevitably waned. Others survived by flogging their wares overseas (www.ambientmusicguide.com 2007).

This vitriolic online review of Australian ambient music hints at both the pervasiveness of the genre, the peculiarities of its marketing system, and the presumed naivety of its purchasers. Our purpose here has not been to assess its artistic merit, but its imagined and real geography. Ambient music, with its links to new age and world music, and its combination of the electronic and the natural, has created a series of generic and particular places, that are said to reflect and induce harmony, and become identified with indigenous Australians. The genre has become exceptionally popular, yet largely distinct from other forms of popular music in the common absence of live performance, its absence from music stores, the absence of critical attention (other than the scathing review cited above, from *The Ambient Music Guide*), and the tourist orientation.

A very particular, essentialized version of Australia coalesces in ambient music—a parallel of sorts to the aural construction of Hawai'i in the 1950s and 1960s as a beachside paradise with friendly Polynesian natives: Australia is a tourist place of harmony, a languid, empty continent (re)inscribed by ambient music and its marketers as home to traditional indigenous peoples, contrasting wilderness landscapes, and animals imbued with spiritual meaning. In a broad sense Australian ambient music has sought to create the backdrop to a marketable, touristic vision of Australia—of wide expanses of brown and red lands, coral reefs, and tropical rainforests—unpopulated and empty, other than the iconic whales, dolphins, and kangaroos (curiously absent from the music, perhaps through not being adequately languid?), and where ancient spirits still inhabit certain iconic sites (such as Uluṟu and Kakadu). Aboriginal people, as the first occupants of Australia, similarly achieve naturalized/essentialized iconic status, with Aboriginal music (meaning didjeridu music, rather than Aboriginal country or hip-hop) said to evoke antiquity and continuity with an idyllic past where harmony with nature reigned. The actual depiction of Aborigines through the music and on the CD covers implies that Aborigines are rooted in, and safely remain in, that past.

Yet, as we have tried to show here, ambient music is also a distinct cultural industry, with its own now sophisticated marketing strategies, distribution mechanisms and impressive sales performances. Two quite different places, Alice Springs and Byron Bay, similar only in their primary tourist orientation, have in different ways drawn ambient music into a tourism soundtrack. In Byron Bay, ambient music has emerged as a distinct, grassroots cultural industry where recording artists market self-made CDs to backpackers in a "new age" destination. Vividly, in Alice Springs, a burgeoning CD market has grown for Aboriginalized recordings, yet largely denying and excluding a contemporary Aboriginal presence (especially given that it is a major center for Aboriginal country music, and rock and roll), both in a broad sense, but also for local Aboriginal musicians (of both traditional and contemporary persuasions), who are excluded from potential economic benefits because of monopolization, national systems for ambient music distribution and retail, and the absence of attempts to broaden tourists' appreciation of Aboriginal music beyond the didjeridu into contemporary styles such as Aboriginal country, hip-hop, and reggae.

What this example reveals for the examination of music in geography more broadly is that music has a particular ability to create and transform places, especially for the tourist gaze/ear. This transformation is simultaneously cultural and economic, representational and material, symbolic and real. Yet music's capacity to transform places cannot be considered incidental, "natural," or neutral. Again as a parallel to Hawai'i (Connell and Gibson 2008), in the case of Australian ambient music there is an inherent bias and distortion in the way the past and present are glamorized, fragments of culture selectively appropriated and marketed, while other realities and cultural expressions remain shrouded from tourist view. Ambient music produces a distinct cultural politics of representation, one with important economic outcomes.

References

Abram, S., Waldren, J.D. and MacLeod, D.V.L. (eds.) (1997), *Tourists and Tourism: Identifying People with Place* (Oxford: Berg).

ACNielsen (2000), *Survey of Indigenous Tourism* (Sydney: Department of Industry, Science and Resources, Northern Territory Tourist Commission, South Australian Tourism Commission, Tourism NSW, Tourism Queensland, Tourism Tasmania, Tourism Victoria.)

Anderson, K. and Jacobs, J. (1997), "From Urban Aborigines to Aboriginality and the City: One Path Through the History of Australian Cultural Geography," *Australian Geographical Studies* 35:1, 12-22.

Atkinson, C.Z. (1997), "Whose New Orleans? Music's Place in the Packaging of New Orleans for Tourism," in S. Abram, J.D. Waldren and D.V.L. MacLeod (eds.).

Carney, G.O. (1998), "Music Geography," *Journal of Cultural Geography* 18, 1-10.

Cohen, S. (2007), *Decline, Renewal and the City in Popular Music Culture: Beyond the Beatles* (Aldershot: Ashgate).

Connell, J. and Gibson, C. (2003), *Sound Tracks: Popular Music, Identity and Place* (London and New York: Routledge).

—— (2004), "Vicarious Journeys: Travels in Music," *Tourism Geographies* 6:1, 2-25.

—— (2004), "World Music: Deterritorialising Place and Identity," *Progress in Human Geography* 28:3, 342-61.

—— (2008), "Exotic Journeys: Music, Record Covers and Vicarious Tourism in Post-war Hawai'i," *Journal of Pacific History* 43:1, 51-75.

Cronon, W. (1996), "The Trouble with Wilderness, Or Getting Back to the Wrong Nature," in W. Cronon (ed.).

Cronon, W. (ed.) (1996), *Uncommon Ground: Rethinking the Human Place in Nature* (New York: W.W. Norton).

Davis, K. (2009), Ken Davis Music <http://www.kendavismusic.com>, accessed 5 March 2009.

Digance, J. (2003), "Pilgrimage at Contested Sites," *Annals of Tourism Research* 30, 143-59.

Dunbar-Hall, P. and Gibson, C. (2004), *Deadly Sounds, Deadly Places: Contemporary Aboriginal Music in Australia* (Sydney and Seattle: UNSW Press and University of Washington Press).

Gibson, C. (1998), "'We Sing Our Home, We Dance Our Land': Indigenous Self-determination and Contemporary Geopolitics in Australian Popular Music," *Environment and Planning D: Society and Space* 16, 163-84.

Gibson, C. and Connell, J. (2003), "Bongo Fury: Tourism, Music and Cultural Economy at Byron Bay, Australia," *Tijdschrift voor Economische en Sociale Geografie* 94:2, 164-87.

—— (2004), "Cultural Industry Production in Remote Places: Indigenous Popular Music in Australia," in D. Power and A.J. Scott (eds.).

—— (2005), *Music and Tourism* (Clevedon, UK and Buffalo, NY: Channel View Press).

—— (2007), "Music, Tourism and the Transformation of Memphis," *Tourism Geographies* 9:2, 160-90.

Hudson, D. (2009), David Hudson: Ancient Voice of the Future <http://www.davidhudson.com.au/>, accessed 5 March 2009.

Indigenous Australia (2009), <http://www.indig.com>, accessed 5 March 2009.

Jackson, P. (1999), "Commodity Cultures: The Traffic in Things," *Transactions, Institute of British Geographers* 24, 95-108.

Janke, T. (1998), *Our Culture, Our Future: Report on Australian Indigenous Cultural and Intellectual Property Rights* (Canberra: Australian Institute for Aboriginal and Torres Strait Islander Studies and Aboriginal and Torres Strait Islander Commission).

Johnson, V. (1996), *Copyrites: Aboriginal Art in the Age of Reproductive Technologies* (Sydney: National Indigenous Arts Advocacy Association and Macquarie University).

Kneafsey, M. (2002), "Sessions and Gigs: Tourism and Traditional Music in North Mayo, Ireland," *Cultural Geographies* 9, 354-8.

Kong, L. (1995), "Popular Music in Geographical Analyses," *Progress in Human Geography* 19, 183-98.

Lanza, J. (1994), *Elevator Music: A Surreal History of Muzak, Easy-listening and Other Moodsong* (London: Quartet Books).

Neuenfeldt, K. (1994), "The Essentialistic, the Exotic, the Equivocal and the Absurd: The Cultural Production and Use of the Didjeridu in World Music," *Perfect Beat* 2:1, 88-104.

—— (1997), "The Didjeridu in the Desert: The Social Relations of an Ethnographic Object Entangled in Culture and Commerce," in K. Neuenfeldt (ed.).

—— (1998), "The Quest for a 'Magical Island': The Convergence of the Didjeridu, Aboriginal Culture, Healing and Cultural Politics in New Age Discourse," *Social Analysis* 42:2, 73-101.

Neuenfeldt, K. (ed.) (1997), *The Didjeridu: from Arnhem Land to Internet* (Sydney: John Libbey and Perfect Beat Publications).

Northern Territory Government (1999), *Foster Partnerships in Aboriginal Development: Foundations for Our Future* (Darwin: Northern Territory Government).

O'Connor, T. (2009), *Tony O'Connor Music* <http://www.tonyoconnor.com.au/ home.htm>, accessed: 5 March 2009.

Our Times (1996), "Making Music," 9, 10-11.

Power, D. and Scott, A.J. (eds.) (2004), *The Cultural Industries and the Production of Culture* (London and New York: Routledge).

Smith, S.J. (1997), "Beyond Geography's Visible Worlds: A Cultural Politics of Music," *Progress in Human Geography* 21, 502-29.

Waitt, G. (1997), "Selling Paradise and Adventure: Representations of Landscape in the Tourist Advertising of Australia," *Australian Geographical Studies* 35, 47-60.

—— (1999), "Naturalizing the 'Primitive': A Critique of Marketing Australia's Indigenous People as 'Hunter–gatherers'," *Tourism Geographies* 1:2, 142-63.

Waitt, G., Figueroa, R. and McGee, L. (2007), "Fissures in the Rock: Rethinking Pride and Shame in the Moral Terrains of Uluru," *Transactions of the Institute of British Geographers* 32, 248-63.

Wang, N. (1999), "Rethinking Authenticity in Tourism Experience," *Annals of Tourism Research* 26, 349-70.

Windsong Records (2003), <http://windsongrecords.com.au>, accessed 6 June 2003.

Discography

Birch, David (1997), *Byron Journey* Compact Disc. Self-released.

Coolangubra (1996), *Spirit Talk* Compact Disc. Celestial Harmonies.

Dargan, Ash (2007), *The Very Best of Ash Dargan* Compact Disc. MSI Music.

Davis, Ken (2003), *Spirit of the Panflute Australia* Compact Disc. Ken Davis Music.

—— (2004), *Australian Atmospheres* Compact Disc. Ken Davis Music.

—— (2004), *Spirit of Sedona* Compact Disc. Ken Davis Music.

—— (2004), *Spirit of the Outback* Compact Disc. Ken Davis Music.

Deal, Steve (Undated), *Spirit of Australia* Compact Disc. Holborne.

—— (Undated), *Didgeridoo Dreaming* Compact Disc. Holborne.

—— (Undated), *Native Birds* Compact Disc. Holborne.

—— (Undated), *Outback* Compact Disc. Holborne.

Denny, Martin (Undated), *Forbidden Island* LP. Liberty.

Dr Didge (1997), *Naked Didge* Compact Disc. Self-released.

Eno, Brian (1978), LP. *Ambient 1: Music for Airports.*

Ganga Giri (Undated), *Primal Pulse* Compact Disc. Self-released.

Holm, Andy (1996), *Blue Mountain Tales* Compact Disc. Primal Harmony Records.

Hudson, David (1996), *Guardians of the Reef* Compact Disc. Indigenous Australia.

—— (1997), *Kuranda* Compact Disc. Indigenous Australia.

—— (1998), *Touching the Sounds of Australia* Compact Disc. Indigenous Australia.

Hudson, David and Friends (1997), *Nessun Dorma "Gari Wunang"* Compact Disc. Indigneous Australia.

Jalalo (1995), *Letting Go* Compact Disc. Self-released.

Kaempfert, Bert (Undated), *A Swingin' Safari* LP. Polydor.

Keech, John (2003), *Pure Meditation* Compact Disc. New World Music.

Melachrino Strings and Orchestra, The (1958), *Moods in Music: Music for Relaxation* LP. RCA.

O'Connor, Tony (Undated), *Awakenings* Compact Disc. Studio Horizon.

—— (Undated), *Hidden Forest* Compact Disc. Studio Horizons.

—— (Undated), *Kakadu* Compact Disc. Studio Horizons.

—— (Undated), *Rainforest Magic* Compact Disc. Studio Horizons.

—— (Undated), *Uluru* Compact Disc. Studio Horizons.

—— (Undated), *Whispering Sea* Compact Disc. Studio Horizons.

—— (Undated), *Wilderness* Compact Disc. Studio Horizons.

—— (Undated), *Windjana – Spirit of the Kimberley* Compact Disc. Studio Horizon.

Rankin, Robert (1994), *Wilderness* Compact Disc. Rankin Publishing.

Rose, Tania (1996), *Coral Sea Dreaming* Compact Disc. Artscope Music.

Si (1998), *Innersense* Compact Disc. Self-released.

Tarshito (1994), *Playing in the Rainbow* Compact Disc. Kittani Music.

Unattributed artist /s (1998), *A Voyage to Australia* Compact Disc. MCPS.

Various artists (Undated), *Flamenco Con Chilli: A Hot Flamenco Blend of Modern Flamenco and Ancient Didg* Compact Disc. Indigenous Australia.

—— (Undated), *Island Warriors* Compact Disc. Indigenous Australia.

—— (Undated), *Rio Didg: A Latino Excursion* Compact Disc. Indigenous Australia.

—— (1999), *Woomera* Compact Disc. Indigenous Australia.

Voices of Gaia (1998), *Save Jabiluka* Compact Disc. Self-released.

Part III
Mapping Musical Texts

Linking landscapes, music, and lyrics can be done through by performing qualitative content analysis of carefully selected songs. In the following two chapters, textual analysis is used to unveil the spatial meaning of places, or put differently, text is re-imagined as maps. Kevin Romig's "A Listener's Mental Map of California" is concerned with the relationship between the music and the audience; how place images, in this case California, are received by listening minds and bodies. Sarah Daynes is also systematically assessing place imagery in music, but in her "A Lesson of Geography, on the Riddim: The Symbolic Topography of Reggae Music" the meaning of place is to be found on a very different level. Reggae music is an internally coherent cultural expression that can tell us much about the worldview of the members of this culture—poor Jamaicans who adhere to the Rastafari religion. This worldview can be understood though semiotic analysis of the spatial features that occupy central positions in reggae songs. Daynes focus on the terms "shantytown," "Mount Zion," and "Africa," which represent fundamental aspects of how reggae music construct a world that in part consists of the harsh, poverty-stricken reality in Jamaica, but also theologically inspired places that straddle the boundary between the real and the imaginary.

Daynes's chapter is an example of a tradition in cultural geography that focuses on collective memories. Reggae not only portrays a particular worldview, but it actively creates and shapes the collective memory of the Jamaican people and the Rastafari movement. This memory is fundamentally spatial in orientation. As representations of "real" geography are not always forthcoming in reggae music, Daynes identifies what she calls the "symbolic topography" of reggae, which creates a semi-imagined, mapable world of reggae. Methodologically, she considers how spaces are articulated in reggae lyrics and the manner in which these spatial entities become interrelated in the texts, thereby teasing out the geographic imagination of reggae.

Daynes limits her analysis to reggae lyrics, which is a concession for the sake of brevity. She rightly acknowledges, however, that such a textual approach presents only a partial read of reggae music (or any form of music). After all, the topic of this book is music, not poetry. Sound also carries meaning to people, which is why music is so important to people. For an analytical integration of music and lyrics, Connell and Gibson's previous chapter on Australian ambient music, and the upcoming chapters on Havana (by John Finn and Chris Lukinbeal) and the Red Hot Chili Peppers (by Michael Pesses) are all excellent reads.

In Kevin Romig's "A Listener's Mental Map of California," lyrics and music from the top of the charts are used to distill the essence of California imagery.

There are few other places in the world that are so defined by popular culture as California. So using music as a lens to untangle the multiple facets of the Golden State is an appropriate approach. Romig shows us how the portrayal of California in music has changed over time and closely mirrors the material realties of the state. To make sense of these discursive trends, Romig has developed six distinct categories, each capturing a specific viewpoint of California as a whole, or specific localities within the state. As the analysis is based on Top 40-style songs, this is music that most people are familiar with, and to some extent, the images conveyed through the songs are also familiar. Romig has identified a strong temporal dimension with easily identifiable representational eras from the optimism of 1950s pop, which projected the spirit of the times onto California, to today's rap which provide a far darker image of the state. Again California becomes not only a distinct spatial entity to be represented, but a microcosm of society's anxieties as a whole. Romig uses an integrated approach where both lyrics and music are ascribed significance. For example, the sounds of surf pop cannot be detached from the lyrics, but forms a coherent whole. In the end, Romig arrives at a mental map of the listener; Daynes, on the other hand, arrives at a mental map of the symbolic topography of reggae, a form of spatialized ideology.

Chapter 6

A Lesson of Geography, on the Riddim: The Symbolic Topography of Reggae Music

Sarah Daynes

At the end of the 1960s, in the studios of Kingston and on the turntables of the sound-systems that traveled over the hills and valleys of Jamaica, a new music started to be played. In a few months, it had won over the Jamaican musical scene; it was reggae music, which has since become the most powerful symbol of the island. In particular, reggae music is extensively used to promote tourism in Jamaica; the official website of the Jamaican tourism office for instance speaks of reggae as being "the heartbeat of our people" (Jamaica Tourist Board 2007). It is widely known that reggae music sings about love and sex, but also conveys a multidimensional message, at once political, economic, social, and religious; this, of course, is best known through the lyrics of Bob Marley, the most famous reggae musician, emblem of Jamaica and international superstar. And yet, very few academic works have focused on an extensive and systematic analysis of reggae lyrics (see, however, King 1995; 2002). In this chapter, I offer an exploration of reggae lyrics based on a corpus of songs (see Discography) elaborated for earlier work on memory in reggae music.

This corpus contains approximately fifty albums, selected from two periods: what is sometimes called the "golden era" of reggae music, from 1973 to 1980, and a more recent period, from 1995 to 1999.[1] These albums were chosen for their consensual significance within the history of reggae music. My analysis is qualitative, in the sense that it did not imply any quantitative counting, for instance, of semantic occurrences; and it is based only on the lyrics, which I transcribed. Hence, I have not taken into account, in my analysis, how the lyrics were sung, the performance, the instrumental part; and neither have I taken into account what the musical critiques or the artists themselves said about the lyrics and their meaning. I have therefore treated reggae lyrics as a text. This approach, of course, is limited to the textual content and structure of the lyrics; this was a voluntary choice, motivated by the desire to take the words of the music in themselves, and on their own ground. Simon Frith once said that "it is not only what they sing, but how they sing it, which determines what a singer means to us and how we

---

1   This second period was, at the time of the elaboration of the corpus, "the present." Indeed this corpus was originally designed in 1999 for my doctoral dissertation (Daynes 2001).

are placed, as audience, in relation to them" (Frith 1987, 97); this article is about "what they sing," not because "how they sing it" is not important, but because "what they sing" also deserves consideration, and is susceptible to be analyzed on its own terms. And indeed, my analysis is not about the reception of the lyrics, but on the lyrics themselves. Hence, I propose here a partial but nonetheless important analysis, which focuses on the words of the music.

Not only are systematic analyses of reggae music seldom; they also tend to focus on religious and socio-political dimensions (Daynes 2001; King 2002). The question of spatiality, when actually raised, is addressed via the religious content of reggae lyrics and the relationship between artists and the Rastafari movement, or via the issue of immigration and the notion of diaspora. Hence within these limits there are interesting analyses already available to us, for instance with the work of Chude-Sokei (1994) on sound-systems, various case studies on reggae music outside of Jamaica (e.g. Garrison 1979 on Great-Britain; Savishinsky 1994 on Senegal; Hansing 2001 on Cuba), or of course the complex relationship to Africa developed by the Rastafari movement and widely expressed in reggae lyrics. Research that focuses on spatiality per se is almost inexistent. And yet, it seems surprising that no scholar has paid attention to the notion of place within reggae music. In this article, I offer a qualitative analysis of reggae lyrics, which provides an understanding of what I have called the symbolic topography of reggae music. My point of departure lies in Maurice Halbwachs's analysis of collective memory and space as found in *The Legendary Topography of the Gospels in the Holy Land* (1942); a process of "memorialization" of space takes place along lines traced by the elaboration of a changing, dynamic collective memory, for which historical accuracy is only a secondary matter; the accuracy here, indeed, is to be found in the logic of symbolic processes. Hence it is not about what things are; it is about what things mean. In other words, I am preoccupied with the meaning attached to places in reggae music, and with the way in which these places are articulated together to form a symbolic topography that relates to exigencies posed by both history and collective memory.

In other words, I am arguing for two crucial points. First, it is not only about places, and how they shape the way in which "the world" is imagined; it is also about *the articulation of these places*, and the structure thereby apparent, which organizes a symbolic topography in the imagination of the group. Collective representations are not just operating at a conscious, straightforward level. In the case of reggae music, I argue that, after a careful reading of the lyrics, a series of "spatial terms" emerge, accompanied by pivotal articulations which structure the specific way in which space is referred to in reggae music. In this article, I offer an analysis of three of these spatial terms: "shantytown," "Mount Zion," and "Africa." The former and latter are "concrete" spatial terms; but Mount Zion is a mythical term, which belongs entirely to the "ideal" sphere ("ideal" not meaning "perfect," but "from the sphere of ideas"). I will analyze two pivotal articulations made between these three terms: Shantytown versus Mount Zion and Mount Zion versus Africa.[2] The term *versus* indicates that there is a dimension of tension between the terms of each dyad; but this tension is not necessarily one of opposition: it is inherent to the articulation between two terms that are both distinct and interrelated. Here, I am not necessarily pointing to a structuralist analysis of binary relationships of terms of opposition (such as the raw and the cooked or the high and the low) as found in the work of Claude Lévi-Strauss. I simply point to articulations that play a fundamental role in structuring the symbolic topography expressed in reggae music, which organizes the world in a specific way on the symbolic level, that is, on the level of collective representations in a Durkheimian sense.[3]

Second, the analysis is not only about space; it is also about *time*. This point will come as a secondary argument within my discussion; but it is nonetheless important, and deserves to be developed further. The way in which the pivotal articulations operate does not relate only to space; it also relates to time: both the places and articulations participate in the elaboration of a collective memory, and also gain meaningfulness from this very collective memory. Hence there is no unilateral relationship of causality here; time builds upon space as much as space builds upon time; and the collective memory is symbolically marked by places as much as the land is symbolically scarred by collective memory. I will return to this point in the conclusion.

The shantytown

It is well known that reggae, since its emergence, came to represent the lower class in Jamaican society and claimed to be the "music of the sufferers," rooted in lower income urban areas; and it is therefore not surprising that the "shantytown"

---

2   Of course, there might be other important articulations if a different corpus of songs was used as a basis of analysis. Additionally, for reasons of space, some terms were left out of analysis here, although they were present in my corpus; it is the case, in particular, of the "diaspora," with references to icons of the struggle for civil rights in the United States, or against racism in Great Britain and Apartheid in South Africa. See for instance Dennis Brown, "Malcolm X," on the album *Visions* (1976).

3   The notion of collective representation is, after 1895, central to Durkheim's reflection on the social quality of thought (Durkheim 1974a; Durkheim 1974b) and on religion (Durkheim 2001). Fundamental to his analysis is the articulation between the internal and the external: collective representations are social facts; they lie precisely in the tension between the collective and the individual; they are both constructed and given, in the sense that they do not belong to the individual and yet do not exist without him; they belong to the social sphere, related and yet not irreducible to the individual sphere. Collective representations are ideas, and are articulated with practices; together, they form social life (Durkheim 1982).

or "ghetto" occupies an important place within reggae lyrics.[4] The most famous is Trench Town, a neighborhood in the city of Kingston in Jamaica, where Bob Marley, Peter Tosh and Bunny Wailer formed the Wailers band in the late 1960s; Trench Town was also the home of several musicians, including Joe Higgs and Alton Ellis.[5] Bob Marley's attachment to Trench Town is reflected in the songs "Trench Town Rock" (1971), "No Woman No Cry" (1974) and "Trenchtown" (1983), which all mention the neighborhood either by its name or by using its district number, "Kingston 12." Twenty years later, in a similar way, in his song "Did You Ever" (1997) Sizzla mentions August Town, another low-income neighborhood in Eastern Kingston (but technically a part of the Parish of St Andrew, which borders the Parish of Kingston), where he lives; he also mentions Riverton City, a Western Kingston neighborhood famous for its dumpster:

> Tell mi if yuh ever walk di streets of the ghetto dem before / fi see who is unsafe from secure / the lifestyle of my people they fail to recognise / do you know what it takes for ghetto people to survive / did you ever walk through August Town at all / fi hear the comments of the people whey a bawl / wonder what it takes for you to realize / how Riverton City ghetto youths dem survive (Sizzla "Did You Ever" 1997).

The shantytown, therefore, is home; indeed, many reggae musicians were born and raised in shanty towns, in what is called in Jamaica "tenement yards" (Bob Marley & The Wailers "No Woman No Cry" 1974; Jacob Miller "Tenement Yard" 1976; Dennis Brown "Down in the Tenement Yard" 1977); but further, the shantytown is also home in a social and political sense. It appears in reggae songs as a symbol for the group, which defines it in opposition to the outside, that is, the Jamaican rich, ruling class. The metaphor for this opposition is organized in spatial terms: the ghetto is an enclave, a neighborhood with well-determined boundaries, within which outsiders do not willingly come. And the "outside" is often referred to as the hill, because affluent neighborhoods in Jamaica are usually situated in the hills, while shantytowns are located in the valleys—where the weather is hotter, the urbanization denser, and the vegetation less luxurious. There is of course something quite socially "perfect" in this spatial organization, in the sense that the dominated, poverty-stricken ghetto dwellers in the valley are looked down upon by those in power, high up in the hills. This is the case in a song by Dennis Brown:

---

4   "Shantytown" is the term used in Jamaica (and elsewhere) for the low income neighborhoods that flourished in the 1960s in urban areas, in particular Kingston and Spanish Town; Desmond Dekker made the term famous with his 1967 hit "007 (Shanty Town)" ("007, 007 / at Ocean Eleven / and now rudeboys have a wail / cause them out a jail / rudeboys cannot fail / cause them must get bail / dem a loot, dem a shoot, dem a wail / a shanty town").

5   See the book *Trenchtown Love* by photographer Patrick Cariou (2004). He also published photos of Jamaican Rastafarians (*Yes Rasta* 2000).

Living in your concrete castle on the hill / you don't know what life is like in the ghetto / living in a two by four with no place to walk around / while you're in a castle all alone … cause life isn't easy in the ghetto / it's hard to keep from getting into troubles … many days we stand in and sun / waiting for the bus that never come / you pass by in your fancy car / wearing a plastic smile and smoking a big cigar (Dennis Brown " Concrete Castle King" 1976).

The ghetto cannot really be ignored; non-ghetto dwellers have to go through it from time to time. But it is only glimpsed at, from within the protective space of the car. The rich live in concrete castles—as opposed to corrugated iron, cramped "two by fours." They drive through the ghetto in superb ignorance, while the poor wait for a bus "that never come." Born in the Jamaican ghettos, reggae echoes their daily life, claims to be the voice of the people, as well as their champion against the oppressors. In "Poor and Clean," Gregory Isaacs affirms that he would "rather be poor and clean than to live rich in corruption," and adds that "the rich man's heaven, heaven, is the poor man's hell."[6] This sentence might be considered the cornerstone of the rhetoric of oppression developed by reggae music since the very beginning of the 1970s: the world consists of rich people (the few) and poor people (the many); Jamaica, still based on a system of a few big land owners for whom many people work ("slave," in the words of Gregory Isaacs), is a deeply unequal society. What appears in reggae music is a description of poverty, but also the idea that poverty is neither a shameful condition nor in the order of things: in reggae lyrics, poverty is sometimes turned into pride, or interpreted as a sign of religious purity; indeed Rastafarian beliefs promise a reversal in the future, as found in several other eschatological Christian religious movements.[7] For Michael Rose, misery is such that there is nowhere to go and no goal to pursue: "Another day of suffering / I woke up this morning with the sky as my roof / I had nowhere to lay my head so the cold ground is my bed … things won't come my way, so it's just another day" (Black Uhuru "Hard Ground" 1977). If Bim Sherman describes, in the song "Down in Jamdown," a world filled with pain, misery and shame, for Bob Marley it is a world where the sun never shines ("Crisis" 1978); in "Concrete Jungle" he adds:

No sun will shine in my day today / the high yellow moon won't come up to play / darkness has covered my life / where is the love to be found? / oh someone tell

---

6　So many years I've been slaving in your factory / never had a chance to talk with the boss / and for so long I've been living in this old community / where no one knew my pain yet I've paid the cost / no one knew my pain no, yet I've paid the cost / but I would rather to live poor and clean / than to live rich in corruption…a rich man's heaven, heaven, is a poor man's hell (Gregory Isaacs "Poor and Clean" 1980).

7　Following the Bible, which states that the stone that has been rejected will become the cornerstone (Psalms 118: 22; Matthew 21: 42; see also Bob Marley & the Wailers "Ride Natty Ride" 1979).

me cause life must be somewhere to be found / instead of concrete jungle (Bob Marley & the Wailers "Concrete Jungle" 1973).

In reggae songs therefore, the shantytown appears both as the place where one was born or raised, or lives in, and as a socio-political referent that symbolizes poverty in a Jamaican society highly based on class (Waters 1989). And, in spatial as much as in socio-political terms, the shantytown is opposed to "the hill," where the rich live. However, in reggae lyrics, the shantytown is also part of another dyad, in articulation with what I have called "Mount Zion." The interesting dimension of this second dyad is its ascription in time as much as in space; and fundamental to it is the idea of hope. While the shantytown and the hill are metaphorically opposed in the present—they describe a current situation—the tension between the shantytown and Mount Zion puts not only two spaces in tension, but also two times: the present and the future.

From the shantytown to the Promised Land

From direct quotes often taken from the Book of Psalms[8] to references made to various passages of the Bible,[9] from the use of grammatical forms and terms characteristic of Biblical English[10] to the opposition between the heathen and the Israelites[11] or Jerusalem and Babylon, reggae music is permeated with the Scriptures. One of the terms that can be found through an endless list of occurrences is Mount Zion, which, accompanied by various related terms such as the Promised Land, Israel, or Jerusalem, forms a multiform category. Mount Zion is a concrete location—a hill near Jerusalem—but in reggae music, it also has to be understood within eschatological beliefs, which concern a mythical narrative of the end of the world, and refers to the New Jerusalem, which will arise after the Armageddon and the Judgment, and where the righteous will live in peace and harmony with God among them (Revelation 21-22). The Judgment of God will punish the sinners and reward the faithful; it is viewed within a rhetoric of oppression that holds the

---

8 For instance, Psalm 23 in Anthony B's "Burn Down Sodom" and "Rumour" (1996), or Psalm 68 in Buju Banton & Garnet Silk's "Complaint" (1995), Barry Brown's "Enter the Kingdom of Zion" (1979), Bob Marley & The Wailers' "Jah Live" (1976) or Anthony B's "Jerusalem" (1997).

9 For instance, to the Book of Daniel in Bob Marley & The Wailers' "Survival" (1979), or to the Book of Proverbs in Buju Banton's "Destiny" (1997), and Bushman's "Live Your Life Right" (1999).

10 Such as the –th past form, and the use of "thou" or "art." For instance "praise yeh Jah" (Sizzla), "He liveth over I and I" (Luciano), "our father who art in Zion" (Buju Banton), and so on.

11 The identification with Israel is visible in the names of reggae bands (The Israelites, Israel Vibration) as much as in songs (see for instance Dennis Brown "Children of Israel" 1977, or Israel Vibration "Walk the Streets of Glory" 1978).

oppressors as the sinners and the oppressed as the faithful. Dennis Brown sings: "Oh what a day, yes oh what a day / when his majesty comes again / to judge everyone / for all the backbiting, cheating and lying / robbing and oppressing the poor / yes he knows they can't take no more … he that causeth the poor to go astray / shall surely pay this day" ("Oh What a Day" 1979).

There is no space here for a deeper analysis of the categories of redemption and hope in the Rastafari movement; it suffices to point that, when evoked in reggae music, the terms mentioned above are ascribed within the religious rhetoric of redemption developed by the Rastafari movement. What matters here is the extra-spatial and extra-temporal quality of this category, which promises another place and another time opposed to the material, human present to which men are confronted. In reggae music, Zion represents an alternative to the shantytown in the sense that it promises a future that will be better than the present, by operating a reversal, following the Scriptures: the stone that the builder refused shall be the head cornerstone (e.g. Psalms 118: 22 and Matthew 21: 42 in Bob Marley & the Wailers "Ride Natty Ride" 1979). Hence there is an articulation between two categories: a present of oppression and a future of redemption. The first category is well symbolized by the shantytown, where the poor live; and the second category, of course, promises liberation. Those today confined to the shantytown, in which life is hard, will in the future be those who inhabit the magnificent Jerusalem. Religious revelation is also articulated to support profane revolution. Hence the term-to-term reversal promised by the eschatology (the oppressed will become the chosen, the oppressors will be punished) also coincides with a term-to-term reversal that can be achieved here on earth, through social-political struggle against poverty and domination. Here again, the shantytown is opposed to "those in power"; for instance in a 1999 song, Anthony B proposes to move Edward Seaga (then Prime Minister) to Tivoli Gardens, one of Kingston's poorest neighborhoods, while the reggae artist Mutabaruka would take his seat at the head of the government: "Tell the Government Jamaica house me come for / cause we want some truth and rights defender / back part a di ghetto give that to Sizzla / all country mi give to Louie Culture / dem a go clean out the heathen wid di Oh Carolina [a famous reggae song] / to Tivoli we a go move Seaga / in a him seat we a go put in Muta" (Anthony B "Me Dem Fraid Of" 1999).

Hence, for the rastas, it is crucial to secure one's place into the City of God, as sung endlessly in reggae music. The Gates of Zion will close upon the believers (Horace Andy & the Sticks "Zion Gate" 1978) and one therefore has to work towards being among them. Interestingly, the New Jerusalem also intersects with the lost origin: hence it is not simply the place of religious redemption, it is also, to some extent, the lost Africa, a parable for unity and peace; it functions as a lost paradise which will be found again, in a parallel with the articulation between Jerusalem and the resurrected New Jerusalem found in the Scriptures. The spiritual path—from evil to good, from oppression to liberation, from the present to future redemption—is articulated in concrete, very spatial terms: one walks to Zion, one leaves the world of evil, symbolized within Rastafarian beliefs by the generic category of "Babylon," to return to the world of God.

> Jah say the time has now come / for all his people to trod Mount Zion / Jah say the time has now come / for I and I to live as one / and yes we are stepping to the land where I and I belong / sure to escape the destruction of Babylon (Hugh Mundell "Jah Say the Time Has Now Come" 1978).

Hence, Zion is an archetypical "elsewhere," the New Jerusalem but also the lost-and-found homeland. The true believers have nothing to fear from the Armageddon; on the contrary, the Judgment is now close and they have to get ready: "Children get ready and trim your lambs / Israelites get steady, no more time to run / cause I say / we moving out of Babylon yeah / we moving to a higher land deh / with the power of the higher man / we moving to Mount Zion" (Luciano "Moving Out of Babylon" 1999). This excerpt refers both to a physical elevation (Mount Zion) and to a spiritual elevation (also linked to the Judeo-Christian tradition of the ascent to heaven by prophets and saints); the Armageddon will be merciless for the sinners, but harmless for the righteous, who are under the protection of God. The former will be thrown in the lake of fire, while the latter will enter the gates of Zion. In "Seek Jah First" (1997), Anthony B tells what will happen after the victory of the Good over Evil: "O what a day that will be / walking down the streets with milk and honey / hand in hand with the majesty / Mount Zion were made for you and me / round the rainbow circle throne / with the man with the crown I and I sat down." The righteous will be saved, and they should therefore trust God unconditionally: "Seek Jah first everything will come off time / Him a di shepherd we a di sheep of his pastures / know the right and truth fullness I-ya / so you pray to the Lord and save ya" (Anthony B "Seek Jah First" 1997).

From Ethiopia to contemporary Africa

It is well known that Rastafarian beliefs give a central place to Ethiopia, mother of the nations, locus of the prophecy, past and present, and of the redemption to come.[12] Ethiopia, hence, represents hope; and, in this dimension, it is often blurred with the category symbolized by Mount Zion (as in Barry Brown's 1977 "No Wicked Shall Enter" or Sizzla's "One Away" 20 years later). In reggae music, Ethiopia occupies an important place, in relationship to the Rastafari movement. The term is found in Anthony B's "Rumour," "Prophecy a Reveal," (1996) and

---

12   See among others Barrett 1988, Chevannes 1994, Johnson-Hill 1994, or Edmonds 2003. The importance of Ethiopia for Rastafari finds in source in Marcus Garvey's words, "Look to Africa, when a black king shall be crowned, for the day of deliverance is at hand," reinterpreted as prophecy upon the coronation of Ras Tafari Makonnen as the Emperor of Ethiopia in 1930, thereby taking the name of Haile Selassie.

"Sunburnt Faces" (1997); in Buju Banton's "Til I'm Laid to Rest" (1995); in Sizzla's "One Away," "Hail Selassie" (1997) and "Strength and Hope" (1999); and in Rod Taylor's "African Kings" (1978). Most occurrences of the term refer to a mythical Ethiopia, religious and mystical, the timeless space that encompasses the beginning and the end, the alpha and the omega.[13]

Interestingly in reggae music, references to Ethiopia are superimposed with references to contemporary Africa; hence Africa is represented, simultaneously, as a mythical place both of origin and redemption, and as a continent of present struggles. One place—the African continent—therefore bears a double temporality: sacred time (whether past or future) and profane time. Until the end of the seventies, Africa was more imagined, remembered and commemorated, than anything else: it then primarily appeared in reggae lyrics as the land of origin, left behind by forced slavery (for instance in "African Race" and "Declaration of Rights" by the Abyssinians, 1976; "Too Long in Slavery" by Culture, 1979; "Jah Can Do It" by Dennis Brown, 1976; or more recently in Luciano's "When Will I Be Home," 1999), or as the locus of future redemption. But progressively, Africa also became a concrete and real continent in reggae music, in particular through two major events extensively relayed in reggae music: the fights for independence, and the struggle against Apartheid.

The independence of colonized African countries took place between the very beginning of the 1960s and the beginning of the 1980s, and strongly marked both the Rastafari movement and reggae music, for two reasons. First, because they echoed the independence of Jamaica, in 1962, and second, because they symbolized the liberation of the African continent, so important to the rastas, and corroborated their faith in God: Africa, free at last, was the proof that the world was changing, and that God was stronger than Evil. In 1979, the independence of Zimbabwe was celebrated in Bob Marley's album *Survival*.

> Every man got the right to decide his own destiny / and in this judgment there is no partiality / so arms in arms, with arms, we fight this little struggle / cause that's the only way we can overcome our little trouble / brother you're right, you're right, you're right, you're so right / we go fight, we'll have to fight, we gonna fight, fight for our rights / natty dread it inna Zimbabwe / set it up inna Zimbabwe / mash it up inna Zimbabwe / Africans a liberate Zimbabwe (Bob Marley & the Wailers "Zimbabwe" 1979).

In this excerpt, Bob Marley addresses both the people of Zimbabwe ("my brother, you're right"), the united group formed by all Africans ("Africans a liberate

---

13 Revelation 22: 13: "I am the Alpha and the Omega, the first and the last, the beginning and the end." Reggae music often mentions King Alpha and Queen Omega, associating men to Alpha and women to Omega, thereby closing the circle: the beginning and the end, the male and the female; completion in God's creation. See for instance Sizzla's "The World" (2000).

Zimbabwe"), and the united group formed by the Africans both within and without the diaspora ("arms in arms we fight" and "we come together"). By celebrating the independence of one country, Zimbabwe, Bob Marley also celebrates the liberation—or the duty of struggle—of the rest of the continent, and a pan-African unity that brings together all the Africans, in and out of Africa. This diasporic and pan-African construction appears very explicitly in another song, "Africa Unite":

> Africa unite, cause we're moving right out of Babylon / and we're going to our fathers land / how good and how pleasant it would be, before God and man / to see the unification of all Africans … so Africa unite, cause the children wanna go home … unite for the benefit of your people / unite for the Africans abroad / unite for the benefit of your children / unite for the Africans a yard (Bob Marley & the Wailers "Africa Unite" 1979).

Marley calls forth the unification of Africa, which also implies the repatriation of the Africans of the diaspora. The African diaspora is therefore imagined within the notion of African unity, as a distinct ("abroad") but nevertheless interconnected group ("children wanna go home"). The physical participation of Bob Marley & the Wailers in the ceremony of independence in Zimbabwe marked the concretization of the symbolic tie between Africa and the diaspora; and the way in which reggae relayed the struggle for independence in the 1970s can be seen as an indicator of the emergence of a "real Africa" in the music, an emergence that was made complete with the South African situation.

Indeed reggae music chronicled the anti-Apartheid struggle, in an almost intimate way: "Ain't gonna sit around and wonder / what to do / cause in South Africa here is apartheid / oh my people … Ain't gonna sit around and wonder / what to do / cause in South Africa here is a fire / on the youth" (Black Uhuru, "No loafing (sit and wonder)," 1980). Black Uhuru suggests unity ("my people"), the worries about the Apartheid regime, and the need for action ("ain't gonna sit around and wonder / what to do"). In 1979, Steel Pulse evokes Steve Biko as a martyr (Steel Pulse "Tribute to the Martyrs" 1979) and tells his story in another song of the same album:

> Blame South African security / a no suicide he wasn't insane / it was not for him to live in Rome / still they wouldn't leave him alone / they provoke him, they arrest him / they took his life away / but can't take him soul / then they drug and ill-treat him, and they beat him / and they claim suicide" (Steel Pulse "Biko's Kindred Lament" 1979).

Biko was an anti-apartheid militant who was arrested and then killed in prison, the case being closed as suicide.[14] Using the metaphor of Rome, Steel Pulse explains

---

14   The case was sent back to court by the Commission of Reconciliation in 1999, and amnesty was denied to the persons implicated in the murder of Steve Biko.

that he was a rebel to the system, here apartheid ("it was not for him to live in Rome"). The identification with Biko, considered an innocent martyr, is marked by a great sense of intimacy, as is apparent in the way his death is described as a personal mourning:

> The night Steve Biko died I cried (and I cried) / Biko, O Steve Biko died still in chains / Biko, O Steve Biko died still in chains / Biko died in chains, moaned for you … I'll never forgive I'll always remember." Source of pain, his death is also considered as an example that will never be forgotten: "not, not only I no / but papa brothers sisters too / him spirit they can't control / him spirit they can't control / cannot be bought nor sold / freedom increase one-hundred fold (Steel Pulse "Biko's Kindred Lament" 1979).

Steve Biko, who "died in chains," is considered a brother and becomes a martyr; he did not die in vain, but on the contrary will serve as a model, as a symbol that reinforces the struggle. Reggae music followed the South African situation over the course of many years; it was described and commented upon, in particular the liberation of Nelson Mandela in 1990 after twenty-seven years of imprisonment, for instance in Barrington Levy's hit "Mandela."[15] From Steve Biko to Nelson Mandela, South Africa was a central theme in reggae music, and this was due to its strong symbolic character. The regime of apartheid indeed represented, in a way, the archetype of domination, pushed to the extreme.

Therefore, there is in reggae music a militant position for African unity and freedom, and against colonization and apartheid. This concrete evocation of the African continent is superimposed upon an imagined, symbolic evocation of Africa—the maternal origin, the Jerusalem of redemption, the eschatological Mount Zion, but also the land of the dynasty of King David and birthplace of the Messiah. Indeed Africa was already massively present in reggae lyrics, but at the end of the seventies it took on a reality that it did not have before: it became a contemporary continent, where milk and honey does not flow, but rather where the struggle against colonialism takes place. This new "reality" of the African continent is especially visible in the way artists began to mention specific African countries, instead of a mythical Africa called "Abyssinia" or "Ethiopia": for instance Bob Marley mentions Mozambique and Angola in "War" (1976); more recently, Buju Banton speaks of Congo, Bostwana, Kenya and Ghana in "Til I'm Laid to Rest" (1995) and of Sudan in "Sudan" (2000), while Anthony B mentions Morocco and Congo in "Conscious Entertainer" (1999); Burning Spear evokes Liberia and Sierra Leone in "Subject in School" (1995) and Luciano refers to Egypt, Morocco, Gambia, Nigeria, Senegal and Tanzania in "When Will I Be Home" (1999).

---

15   Mandela, Mandela, Mandela you're free / they tried to keep a good man down but they can't / lick a shot if you love Mandela / Mandela you're free (Barrington Levy "Mandela" 1992).

These "two Africas" are neither in contradiction nor in conflict with each other, but permanently and simultaneously complementary. The difficulties faced by many African countries do not endanger what Africa represents at a symbolic level; Africa continues to provide hope and still constitutes the origin as well as the redemption to be attained. It is sometimes difficult, when Africa is evoked in reggae music, to know exactly "which Africa" is spoken of, because the two levels are not completely distinct, but intermingled. However, this must not be considered as a contradiction: indeed Africa is both a reality and a symbol, a continent in pain and the land of redemption, different countries (which are sometimes at war with each other) and a united Ethiopia which bears redemption in unity.

Meaning, space, time: The symbolic topography of reggae music

This short glimpse at the spatial articulations present in reggae songs—between the shantytown, a mythical Africa and a real Africa—provides an interesting understanding of what I have called symbolic topography. The world, indeed, is not only a map, which shows the surface of the land and seas on earth using Cartesian coordinates and one of many available projections. The history of cartography shows that the world is represented not only by what is known about places, but also by why places matters.[16] To the problems raised by the representation of something curved (the earth) on something flat (the map) are added problems related to meaning, to the symbolic importance of specific parts of the land and seas, and the connections between them. Groups (and individuals) make their own maps of the world, dependent on meaning, which coexist with consensual representations of the earth (such as the Mercator projection with its built-in inaccuracies). One example would be the cartogram, which transforms the earth's surface according to population or any other indicator, but more specifically, I am thinking about the famous surrealist map of the world published in 1929 in a volume of the Belgian magazine *Variétés* dedicated to surrealism. Called "Le monde au temps des surréalistes," it draws the world according to the *symbolic importance* of places for the surrealist movement. Europe is minuscule, the United-States absent, Russia is immense, Easter Island is as big as South America, and Oceania lies, dominant, at the center of the world. The map represents the world based on its utopian importance (the surrealists were then sympathetic with the Russian Revolution), its artistic influence (hence the central space occupied by Oceania), and its dream quality. What matters is not the land as it is; rather, the land as it is imagined, thought, and represented.

Such a map could be drawn for reggae music as well. The symbolic topography of reggae music gives a central place to Jamaica—an island viewed through the fundamental distinction made between the shantytown and the hill—and to a

---

16   For instance, Saint Isidore's map of the world (1475) includes the three known continents (Europe, Asia and Africa) within a structure marked by Christian thought.

multileveled Africa, both locus of the profane origin and religious redemption to come, and of a contemporary Africa. Israel and in particular Jerusalem and Mount Zion form one more pole, not in a contemporary sense, but significant on the level of the religious past and future. And to a certain extent Israel and Africa are blurred together, depending on the dimensions they both take at a given moment. The Atlantic Ocean lies in between these three poles, a space of connection between exile and the land of origin, a space marked by the slave trade but which also bears the promise for repatriation.

But that is not all; the meaning of reggae is not just a question of space. Time superimposes the places, it nourishes them with meaning. Jamaica is Egypt; Africa is the land of Canaan. The Atlantic Ocean is reddened by the blood of the slave trade and equated with the Red Sea, crossed by the Israelites on their way to redemption.

> We gonna walk the streets of glory / Babylonian gonna run and try to get there before us / it's like the time in Egypt, when Egypt chased after Israel through the Red Sea / then it's like right now dem a chase after Israel through the bloody sea" ("Walk the Streets of Glory" 1978).

Chased across the Red Sea as they escaped to Canaan in the past, the children of Israel are chased again today as they attempt to escape exile, this time across another sea still red ("the bloody sea") from the blood shed by the slave trade. Contemporary struggles are equated on one hand with the religious past (here the identification with Israel in bondage at the hands of Pharaoh works superbly, in relationship with the African slaves brought to slavery to Jamaica) and on the other hand with the religious future (as they are interpreted as "signs" or "steps" within the eschatological narrative—for instance the anti-Apartheid struggle). The arrangement of space, the mapping of the world in thought, takes on a thick symbolic texture that cannot be detached from a complex arrangement of time. Africa is both past, present and future; it is both the land of the offspring of King David, locus of the redemption, and the maternal origin from which the slaves were taken away; it is also the Africa of Apartheid (itself viewed both as a socio-political event, as a symbol of past religious history, and as an eschatological sign) or of the genocide in Sudan; an Africa both unified in the religious referent "Ethiopia" and made up of distinct, contemporary countries. Hence space is given texture by time—and here, in the case of reggae music, by a conception of time that articulates, in a complex manner, the profane and the sacred, historical time and eschatological time. Space is not only about land; it is also about meaning.

References

Barrett, L.E. (1988 [1976]), *The Rastafarians* (Boston: Beacon Press).
Cariou, P. (2000), *Yes Rasta* (New York: Powerhouse Books).

—— (2004), *Trenchtown Love* (Paris: 779 Editions).

Chevannes, B. (1994), *Rastafari: Roots and Ideology* (Syracuse, NY: University Press).

Chude-Sokei, L. (1994), "Post-nationalist Geographies: Rasta, Ragga, and Reinventing Africa," *African Arts* 27:4, 80-84.

Daynes, S. (2001), *Le mouvement rastafari: Mémoire, musique et religion* (Paris: Ecole des Hautes Etudes en Sciences Sociales).

Durkheim, E. (1974), *Sociology and Philosophy* (New York: The Free Press).

Durkheim, E. (1974a [1898]), "Individual and Collective Representations," in E. Durkheim.

—— (1974b [1911]), "Value Judgments and Judgments of Reality," in E. Durkheim.

—— (2001 [1912]), *The Elementary Forms of Religious Life* (New York: Free Press).

—— (1982 [1895]), *The Rules of Sociological Method* (New York: Free Press).

Edmonds, E.B. (2003), *Rastafari: From Religious Outcasts to Culture Bearers* (Oxford: University Press).

Frith, S. (1987), "Why do Songs Have Words?" in A.L. White (ed.).

Garrison, L. (1979), *Black Youth, Rastafarianism and the Identity Crisis in Britain* (London: Acert Project Publication).

Halbwachs, M. (1942), *La topographie légendaire des Evangiles en Terre Sainte: Etude de mémoire collective* (Paris: Puf). [Published in English in *On Collective Memory*, Chicago: The University of Chicago Press, 1992.]

Hansing, K. (2001), "Rasta, Race and Revolution: Transnational Connections in Socialist Cuba," *Journal of Ethnic and Migration Studies* 27:4, 733-47.

Jamaica Tourist Board, <www.visitjamaica.com>, accessed February 2007.

Johnson-Hill, J.A. (1994), *I-sight, the World of Rastafari: An Interpretative Sociological Account of Rastafarian Ethics* (Methuen/London: ATLA/The Scarecrow Press).

King, S. and Jensen, R. (1995), "Bob Marley's Redemption Song: The Rhetoric of Reggae and Rastafari," *Journal of Popular Culture* 29:3, 17-36.

King, S. (2002), *Reggae, Rastafari, and the Rhetoric of Social Control* (Jackson: University Press of Mississippi).

Savishinsky, N. (1994), "Transnational Popular Culture and the Global Spread of the Jamaican Rastafarian Movement," *New West Indian Guide* 68:3-4, 259-81.

Variétés (1929, June), "*Le surréalisme en 1929*." Bruxelles, *Variétés* editions, volume hors-série.

Waters, A. (1989 [1985]), *Race, Class, and Political Symbols: Rastafari and Reggae in Jamaican Politics* (New Brunswick/ London: Transaction Publishers).

White, A.L. (ed.) (1987), *Lost in Music: Culture, Style and the Musical Event* (London: Routledge).

Discography

Abyssinians (1976), *Satta Massagana.*
Anthony B. (1997), *Universal Struggle.*
—— (1999), *Seven Seals.*
Banton, Buju (1995), *Til Shiloh.*
—— (1997), *Inna Heights.*
Black Uhuru (1977), *Love Crisis*. Remixed and re-edited in 1981 as *Black Sounds of Freedom.*
Brown, Dennis (1976), *Visions.*
—— (1977), *Wolves and Leopards.*
—— (1979), *Joseph's Coat of Many Colours.*
Burning Spear (1995) *Rasta Business.*
Bushman (1999), *Total Commitment.*
Culture (1979), *International Herb.*
Isaacs, Gregory (1980), *Once Ago.*
Israel Vibration (1978), *The Same Song.*
Marley, Bob, and the Wailers (1971), "Trenchtown Rock." In *African Herbsman*, 1974.
—— (1973), *Catch a Fire.*
—— (1974), *Natty Dread.*
—— (1976), "Jah Live." Single, recorded 1975 after the death of Haile Selassie.
—— (1978), *Kaya.*
—— (1979), *Survival.*
Miller, Jacob (1976), *Tenement Yard.*
Mundell, Hugh (1978), "Jah Says the Time Has Now Come." In *Classic Rockers*, 1995.
Sherman, Bim (1999), *Love Forever 1974-1979.*
Sizzla (1997), *Praise Yeh Jah.*
Steel Pulse (1979), *Tribute to the Martyrs.*
Various (1999), *Jet Star Reggae Hits, Volume 25.*
—— (2004), *Jah Love Rockers 1975-1980.*

Chapter 7
A Listener's Mental Map of California

Kevin Romig

"California—the land of fruit and nuts" (Texas folklore)

California is a place encompassing diverse terrain, multiple cultures, a large population, and various meanings within popular culture. As a state historically associated with a gold rush, many have long thought of California in a positive way as a place to reverse or accentuate one's fortunes and make a new start. Its moderate climate and lifestyle have been sought after. To others, it is a far-away, broken place imbued by earthquakes, fires, crime, and eccentric people. This chapter seeks to tie together many of these various meanings of how California is viewed as a *place* within the framework of popular music.

Geographers have frequently engaged in the study of popular culture, and often these studies have encompassed themes about popular music. One of the important forerunners in this endeavor was Larry Ford's 1971 article on the origin and diffusion of rock and roll music. That piece has been an important part of cultural geography education throughout the world. Recently, more critical studies of popular culture have influenced the way scholars approach studies on popular music and how it is performed and received. Advances in semiotics have led to a better understanding of how signs and images are received by the public. This study is keenly interested in how people's images of places that stem from popular music might influence their perceptions of place. While visual images are possibly more poignant in spatial awareness, they allow for a more narrow interpretation of imagery. Whereas music lyrics are more open-ended and are perhaps more ubiquitous as we are constantly bombarded with music as we drive, shop, attend sporting events, and eat at restaurants. Popular music intendeds to mediate our everyday lives, but how is meaning imbued in this "soundtrack" of our lives?

California is widely appreciated as a site for significant image creation through the television and film industries clustered in Hollywood. As sound technology increased, film directors and producers desired better access to popular music and musicians to enhance sound tracks to their films. As musicians flocked to this emerging cultural hearth, their geographical situation and their journey became an important element in their song writing. Whether or not the tunes were part of films, California's music industry began flourishing throughout the late twentieth century. As Los Angeles (and more specifically Hollywood) emerged as musical node, more talent kept coming, and through inertia, Los Angeles became not only a film capital but also an important location for popular music production. Through time, popular music resources dispersed across the state and different locales became nodes in a broad music hierarchy and are widely referenced in song lyrics. These lyrics and their associated music styles are the focus of this chapter. Specifically, how has the state of California been represented in popular music lyrics, how has that image changed over time, and what is the significance of those changes in how people view place in music?

Key literature

The literature within music geography rapidly increased along with comments about the socio-cultural components of the production of music. Many of the early writings focused on issues of concentration and diffusion whether it was based on electronic diffusion (Ford 1971; Horseley 1977), concentrations of birthplace locations of musicians (Carney 1974, 1979) or the diffusion of musical styles across the landscape (Arkell 1991; Kuhlken and Sexton 1991). During the 1990s, music geography began taking on a more critical approach favoring theoretical narratives over spatial analysis partly in response to Lily Kong's criticism of music geography (Kong 1995). Literature focused more on interplay between institutions, agency, and their discursive practices in the creation of music places (Leyshon 1998). Place, imagery, and identity became the focus as seen in Saunders's (1993) discussion on the African-American ghetto and Bowen's (1997) description of cultural fascination with Jimmy Buffet's Margaritaville.

   Where some geography literature has been somewhat weak is in understanding the relationships between place, imagery, and expression as a reflexive process between music and listeners. In describing music, the trend is to be guarded, vague, and general so as to not offend or be viewed as pedantic within an increasingly postmodern academia (Kramer 2003). However, music takes on an explicit representational function and has enormous capacity to elicit feelings in listeners (Walton 1997). Lyrics are literary works occasionally highlighting geographical themes that produce fictional (or real) worlds a listener may tap into. This "new musicology" realm has not been pursued enough in geography, but it is legitimated with the understanding that music is a socially constructed process between the listener and the musician. Lyrics carry meaning to recipients and influence their perceptions (Shepherd 1991). While this process can only generalize perceptions and makes certain assumptions on how symbols and signals are received, the information and research may also enhance our way of engaging research on themes of place and how music engenders people's sense of place.

Methods

Content analysis is defined as a method that systematically examines the content of written material (Krippendorf 1980). A straightforward content analysis enables the researcher to identify the number of references to the state in popular music and find trends between words in music lyrics. By adding semiotics to the content analysis, not only can the importance of key terms be quantified, but the meanings of the material can be better understood in terms of how it is received by a listener (Shepherd 1991). The historical and socio-political context of the music is an essential element of not only what they author meant, but how the messages were received. The music about California, lush with symbolic terms and phrases, influences new and distinct ways of speaking, dress, and image processing because of its popularity and thorough electronic and digital diffusion.

Because of the popularity of California and parts thereof in popular music, a few important delimitations guide this study. First, a song is considered "popular" if it charted in the top twenty places of the *Billboard* charts from 1955 through 1999. This is not intended to diminish the influence music that did not make the *Billboard* magazine charts, but it scales this enterprise to a more manageable level. While the Top 20 songs are the focus here, many other less popular but perhaps more important songs also reveal important imagery about California. While the *Billboard* Top 40 charts are not a perfect measure, they are based on radio airplay and sales and are a fine surrogate for popularity (Whitburn 2000). A second imperative delimitation is how a song about California is defined. For this paper, a song needs to be explicitly about California or parts thereof and the overall message of the song implies something about California or the Californian lifestyle. To gather these songs, the *Green Book of Songs by Subject* (Green 1995), a volume listing different songs categorized by subject was utilized. I sought California and cities within California as subject headings to search, but certain tunes were eliminated because of a lack of a central focus on California. Songs like Guns and Roses "Paradise City" while being about Los Angeles in the minds of the group and its immediate fans, does not include specific references to the state that a casual listener could interpret. Therefore, a song needs to be overtly about California to be included in this study.

To better illuminate recurring themes, the songs are organized into six different categories based on their narrative, which includes both the musical melody and lyrics. There is a temporal progression to these themed categories often based on broad common beliefs and socio-economic conditions in the state. As the themes change, so do the common beliefs about California in the minds of listeners. Musically, California in the 1990s is a very different place than the 1950s. The six categories are introduced in their historical progression and the themes that will be pursued are titled: Escapism and Far-away Romance, Exuberance in the Surf and Drag Era, Experimentalism and the Summer of Love, Downhearted and Downtrodden in Hollywood, Hedonism, Intoxication and the Voyeur, and Masculinity and the Gangsta's Paradise (Table 7.1).

Escapism and Far-away Romance

In 1950s America, California had become a popular setting for films, but it was not yet considered an important cultural hearth. As television and jet air travel became

Table 7.1   Songs charting in the Top 20 and their categorical themes

Song	Artist	Position on Chart	Year
Escapism and Far-away Romance			
City of Angels	Highlights	19	1956
26 Miles	Four Preps	2	1959
I Left My Heart in San Francisco	Tony Bennett	19	1962
California Dreamin'	Mamas and Papas	4	1966
Promised Land	Elvis Presley	14	1974
Exuberance in the Surf and Drag Era			
Surfin' USA	Beach Boys	3	1963
The Little Old Lady (From Pasadena)	Jan & Dean	3	1964
California Sun	The Rivieras	5	1964
California Girls	Beach Boys	3	1965
California Nights	Lesley Gore	16	1967
Experimentalism and the Summer of Love			
San Francisco	Scott McKenzie	4	1967
San Franciscan Nights	Eric Burden and the Animals	9	1967
Monterrey	Eric Burden and the Animals	15	1967
Dock of the Bay	Otis Redding	1	1968

Song	Artist	Position	Year
We Built this City	Starship	1	1985
Downhearted and Downtrodden in Hollywood			
Do You Know the Way to San Jose?	Dionne Warwick	10	1968
It Never Rains in Southern California	Albert Hammond	5	1972
Country Boy	Glen Campbell	11	1976
Hollywood Nights	Bob Seger & The Silver Bullet Band	12	1978
Say Goodbye to Hollywood	Billy Joel	17	1981
Hedonism, Partying and the Voyeur			
Hollywood Swinging	Kool and the Gang	6	1974
Hotel California	Eagles	1	1977
Free Fallin'	Tom Petty	7	1990
All I Wanna Do	Sheryl Crow	2	1994
Masculinity and the Gangsta's Paradise			
Nuthin but a "G" Thang	Dr. Dre	2	1993
Wit Dre Day	Dr. Dre featuring Snoop Dogg	8	1993
California Love	2 Pac with Dr. Dre	6	1996

Source: Whitburn (1999). Reprinted by permission of Nielsen Business Media Inc.

more prevalent, California became more embedded in the mental map of many Americans and people around the world. The popular songs of this era reflect a sultry setting as seen through the foggy Golden Gate Bridge in San Francisco, the safe warmth in LA, and the "tropical" hideaway of Catalina Island (Figure 7.1). This tropical setting also produces a hazy depiction of California as a place as the imagery is somewhat fuzzy and not overly detailed. In "City of Angels" the lyricist describes a romantic story within earshot of the bells of San Pedro where two young lovers begin an amorous relationship. In a similar way, Tony Bennett croons about, "Leaving his heart in San Francisco" with his love that, "waits above the blue and windy sea." The Four Preps describe the landscape of Catalina Island as, "A tropical heaven out in the ocean covered with trees and girls." The recurring theme is one of escapism and once you get to California, you will find a tropical and romantic place.

The other two songs within this theme are more comparative showing the spatial relationships between California and somewhere else. In "Promised Land" (written and performed by Chuck Berry nearly twenty years before Elvis had a hit with this song) the singer describes the journey to the "Golden State" where California is the Promised Land for a poor boy from the American South. Numerous travel difficulties do not stop the main character from arriving in Los Angeles and calling home to tell the "folks back home" that he made it to the wonderful destination. In the Mamas and the Papas hit "California Dreamin'" a stark comparison is drawn between a cold, gray, barren unspecified winter location and the sunny warmth of Los Angeles. The main character desires to leave to go to California, but they are held back by torn emotions and a hint of a devolving amorous relationship. The phrase California Dreamin' is widely used in considering another locale with the mild conditions in California. While the imagery is not overly vivid, the message is clear in that California is better than where you are. In general, the songs within this category lay an important groundwork for how California is viewed throughout much of the twentieth century. The remaining storylines in popular music simply accentuate these stereotypes as a foil to reveal a different picture or take substantially graphic efforts to counter these popular beliefs. Through time, imagery will become more colorful, but this music stands as the foundation for representations to come.

Exuberance in the Surf and Drag Era

Pop musicologists agree that the role of youth and the emergent teenage market was incredibly important in the success of the recorded music industry (Wicke 1990). Due to great industrial expansion, disposable incomes rose and the teenager became an emergent social group willing and able to afford to purchase records and spend spare time listening to the radio. Innovative artists like Chuck Berry and Buddy Holly reacted to this by targeting their lyrics and melodies to this social group. This next important era or genre in California popular music highlights a younger, more exuberant, detailed picture of two different Southern California

**Figure 7.1   Map of California and notable locations in popular music
and culture**

countercultural lifestyles—the surf scene and beach life as well as the drag racing car culture popular in the inland valleys of the Southland. The Beach Boys and Jan and Dean were instrumental in elevating the popularity of this music as well as reflecting the youth culture of California for the rest of the world. The members of these groups were from Southern California adding more credibility and a "reality" to the enterprise. As this musical style evolved, groups like the Rivieras (from South Bend, Indiana) attempted to tag onto the success of this movement. The melodies often comprise of tight harmonies backed by throbbing reverberated guitar sounds meant to symbolize the swashing ocean.

The first major hit of this genre clearly signifying California as a center of surfing culture was "Surfin' USA" by the Beach Boys in 1963. Utilizing an already popular tune 'Johnny B. Goode" as the melody, this song describes the look of a surfer (sandals and a bushy blonde hairdo) the life of a surfer (to go on "safari" once school is completed in June) and the popular hangouts for surfing (Pacific Palisades, Manhattan Beach, San Onofre, and so on). While these minor beach cities may have been somewhat obscure to the non-Californian in 1963, the Beach Boys solidify California as the hearth of surfing with their introductory line, "If everybody had an Ocean, across the USA, then everybody would be surfing, like California." The Beach Boys two years later penned a hit describing the attractive qualities of "California Girls" as being tanned, and dressed in a French bikini. They are deemed superior to attractive woman across the USA as well as in "this great big world."

Jan Berry and Dean Torrance together formed the vocal group Jan and Dean. They were large figures in this music genre with hits such as "Surf City" and "Drag City" in 1963 which highlighted the competing countercultural movements. While these songs generally shed light on California, their 1964 hit, "The Little Old Lady (from Pasadena)" illuminates the San Gabriel Valley and the popularity of drag racing (Figure 7.1). The song's main character is a typical suburban grandmother who happens to own and drive a "shiny red Super-stock Dodge" quite passionately on the roadways of Pasadena. Lingo such as "blowers" and "shut down" infuse slang to add connections to the countercultural activities. One might be left with the imagery of an older female blasting down the Tournament of Roses Parade route leaving boisterous young adolescent males in the dust.

The two other Top 20 hits from this era specifically mentioning California or parts thereof highlight the popularity and adaptability/portability of this genre. Lesley Gore's hit "California Nights" was a bit of a reach for the singer who became famous during the Philadelphia schlock era (a time described by musicologists as when rock and roll was overtaken by large corporate conglomerates to soften the image of the music through well-dressed and groomed artists) of the early 1960s. It does, however, effectively tie together the romance of the first theme with beach imagery of this genre as the main character walks hand-in-hand with her boyfriend along the beach. "California Sun" by the Rivieras is an enjoyable song as the lyrics follow a dream of having fun partying along the beach and dancing in the warm California sun. The Surf and Drag movement was incredibly important in establishing California as a legitimate cultural and music hearth. Its immense popularity during the early and mid-1960s provided music listeners an alternative to the concurrently popular British Invasion. While the music does not specifically challenge the escapist romanticism of the earlier period, it does provide more detail into the youth consumer culture of Southern California and help export Californian lingo and pastimes to other parts of the world.

Experimentalism and the Summer of Love

While the surf and drag music portrayed blithe and amiable teens along the coastline and valleys concerned with waves and autos in the mid-1960s, a movement of confrontation and resistance was gaining popularity and prominence in Northern California. The Free Speech Movement (FSM) on the campus of the

University of California in Berkeley (Figure 7.1) empowered students to take a stand for civil rights and against "faceless bureaucratic machines" and the intensification of fighting in Vietnam. Other young people of the time responded to this by seeking alternative styles and approaches to life in contrast to the dull and flavorless popular society (Faragher 1994). A counter-culture began to develop in San Francisco, centered in the Haight-Ashbury District, organizing and strengthening people of dissenting popular viewpoints. The culture was centered on ideals of communal living, anti-consumerism, simpler ways of life, and a deep global awareness (Rycroft 1998). Artistic elements including poetry and music were key ingredients to this subculture. Another important aspect of the San Francisco subculture was the use of lysergic acid diethylamide (LSD) which added a layer of disapproval (from broad society) to the hippie subculture. Much of the music developed in this scene intended to re-create the "trips" of the mind-altering hallucinogen.

Scott McKenzie's folk anthem "San Francisco (Be Sure to Wear Some Flowers in Your Hair)" was a symbolic invitation to and portrayal of the Summer of Love (1967) in San Francisco. The wearing of flowers in your hair signified peace and the beauty of Mother Earth, and McKenzie implores the "generation with a new explanation" to congregate harmoniously in San Francisco to register their displeasure with mainstream society. Likewise, Eric Burdon and The Animals' hit "San Franciscan Nights" (1967) also hints at the welcoming and "warm" reception all will receive in San Francisco during the Summer of Love. In the song, "walls move, and minds do too" alluding to the hallucinogenic nature of LSD. A police officer's face that is "filled with hate" symbolizes the tension between the law and the protestors. The Animals had another important hit later with 'Monterrey" which describes the events at the Monterrey Pop Festival in 1967 believed to be the musical pinnacle of the Summer of Love with performances by Jefferson Airplane, The Who, Grateful Dead, and Jimi Hendrix (Figure 7.1). Monterrey was the catalyst for future musical festivals such as Woodstock and Altamont and the rapid and strong musical background to the lyrics capture the excitement and exuberance associated with the music festival.

The remaining two songs associated with this era are not necessarily as poignantly emblematic of the Summer of Love, but either project images of the welcoming nature of San Francisco or reflect back on the San Francisco scene one generation removed. Otis Redding's hit (1968) "Dock of the Bay" describes the relaxed life of a person who has relocated from Georgia to the San Francisco Bay Area. After the move, however, his life still remains lonely and unfulfilled. Starship (a re-tooled version of Jefferson Airplane) released, "We Built this City" in 1985 which laments the changes in the San Francisco music scene and how the place had returned to a more corporate landscape akin to other major American cities. It is a peppy tribute to the work of themselves and other musicians of the 1960s who built the city on rock and roll, but the music itself is not evocative of the bluesy, earthy rock of the 1960s but is a pop-like ballad characteristic of 1980s music.

The San Francisco scene was truly important in offering a differing view of life in California in opposition to the exuberance and romance of Southern California. This was not only important musically, but also socially as the music encouraged young people to become more aware of political issues, many of them violent, happening within the United States and caused by the US abroad. While overall the message of the peace and love movement was related to proactive, positive change, the image of shaggy hippies using marijuana and other hallucinogens decrying mainstream society began an escalation of negative stereotyping of California and its people.

Downhearted and Downtrodden in Hollywood

The economy of California could not keep up its rapid pace of the 1950s and as the economy began slowing, local residents' hopes and dreams about California life began tarnishing. California's unemployment rate jumped from 5.8 percent in 1960 to 7.3 percent in 1970 and 9.9 percent in 1982 according to the California Almanac. Music about California from this era tend to be softer more "poppy" hits that tell stories of trials and tribulation in the fast-paced, phony, and unnatural landscape of Los Angeles. Songs of this era challenge the popular discourse of California being the "Promised Land" and offer an alternative vision of depression, disappointment and failure. The melodies take on a secondary role to the poignant, smart lyrics that depict a vivid picture of remorse and sadness. The negativity begins in 1968 with Dionne Warwick's hit that asks, "Do You Know the Way to San Jose?" Here the main character considers a relocation from Los Angeles's smog, empty promises, and traffic-plagued streets to the clean, quiet, friendly locale of San Jose. In a similar vein, Glen Campbell's 1976 song "Country Boy (You've Got your Feet in LA)" tells a story of a man whose dream was to make it as a big star in California, but he still yearningly longs for the peaceful serenity of the intimate music scene of Tennessee. He feels like a "sell-out" and has lost control of his life. Los Angeles is not home and the main character comes off as unnerved.

Analogous to the theme in Glen Campbell's hit, "Hollywood Nights" is Bob Seger's ode to the overwhelmed Midwestern boy who has fallen for an experienced California beauty. In this era, California is rarely depicted as home, but rather a place of lost identities and confused times. The lyric, "he was too far from home" is repeated in "Hollywood Nights" to offer the listener a lens into the boy's troubled heart. At the end of the song, we are left with the scene of the main character waking up alone as the girlfriend has left for greener pastures with someone else, leading him to wonder "if he could ever go home." Ephemeral relationships are the main storyline to the 1981 charting hit "Say Goodbye to Hollywood" by Billy Joel. As the main character is preparing to leave Los Angeles, he reminisces about how his friends have turned their back on him for saying one thing out-of-line. To him, "Life is just a series of hellos and goodbyes" and he is saying goodbye to the Hollywood life and its shallow people.

"It Never Rains in Southern California" uses anecdotal sayings about the moderate climate of California as a metaphor for unrealized dreams and hard luck in an area that promises good fortune and stardom. The song does not categorize an "El Nino" event; rather, bad luck pours on the main character who is, "out of luck, out of my head, out of self respect, and out of bread (money)." He further describes different opportunities that do not materialize in television and the film industry. When he comes across a familiar face from his hometown he pleads with them to not let his family know of his hardships and shame. The basic idea is that in Hollywood, expect your best dreams to become your worst nightmare or that things are not as they seem. The melancholic attitude of California in this era is historically significant as this is a time of the end of American innocence. Numerous assassinations of famed leaders, the Vietnam War, Watergate, and inflation during the shift from Fordist to Post-Fordist production subdued the overall mood of the nation which is a likely reason these songs reached a high level of popularity. This is also the era when photochemical smog became a large problem in Southern California as urbanization, industrialization, and physiographic forces combined to create unhealthy air in the Los Angeles basin and valleys. The future seemed more dystopic than utopic. These songs struck an important cultural chord with many who struggled with what the future would hold within a society that was less trustworthy of their friends and fellow citizens.

Hedonism, Intoxication, and the Voyeur

In many ways, this theme or genre is a "catch-all" for songs that do not fit as neatly into a category of music, yet upon closer inspection two important themes emerge from these four songs: intoxication or blurred sense of self, and general partying or celebrating. "Hollywood Swinging" celebrates the glitz and glamour of Hollywood as it is basically the story of how Kool and the Gang became successful stars after their move to Hollywood. It is a peppy and jovial tune that is in stark contrast to many of the other portrayals of California at the time. "Hotel California" is perhaps the most popular song associated with the Golden State, but this association is very weak at best. While it is likely that the Hotel California is somewhere within the state, all we are told about the surrounding area is that it is in a desert location with a nearby mission bell. The main character is drawn to an enticing "woman of the evening" who works at the hotel and spends a romantic evening dancing, drinking wine, and having sex. Hedonism is a key ingredient of "Hotel California" as the partners give into temptation within a room decorated with a mirrored ceiling and champagne on ice for consumption leading to the physical event. At the end of the song, the main character feels badly about the escapade and looks to leave, but is reminded that he can go but he will not be able to forget the passionate night spent with his companion.

As the 1990s arrive and Generation X is the target audience for popular music, many of the themes tend to highlight popular feelings of this social group: cynicism, skepticism, and alienation. The Cold War had come to an end and the economy in much of California was feeling negative affects of defense spending cut backs. As capitalism reigned supreme of the world's economic system, Generation Xers began to notice stark differences in class and social level. Two songs from this era illuminate these feelings as somewhat intoxicated narrators reflect on the more perfect world that seems to surround them.

"Free Fallin'" is a song of contrasts. The story is told from a male perspective as he reminisces about a girl who seems to be perfect in many ways. She loves her "Mama, loves Jesus and America too." The main character continually defines himself as a "bad boy" for breaking her heart. The romance between the two failed and the boy is interested in "leaving this world for a while" likely using drugs of some sort to dull his pain. A secondary contrast concerns the location of the San Fernando Valley, which is thought of largely as polished suburbia in Los Angeles, but in and this story it is the setting of teenage romantic dystopia. Harkening back to some of the music of the San Francisco scene, the pulsating guitar riffs toward the end of the melody imply the pounding of one's heart while under the influence of drugs. Likewise, Sheryl Crow's pop hit "All I Wanna Do" is also set in Southern California, but in a different situation—a bar along Santa Monica Boulevard. The female character is spending time in the tavern observing both the bar patrons as well as, "the good people of the world" washing their cars at a nearby carwash. The character is clearly poking fun of the status quo and quips that she is more interested in having fun than finding an unfulfilling job to keep up with societal norms. It is an anthem for female empowerment and seeking alternatives to the typical life of a tanned California Girl. These songs strike a chord with typical feelings associated with Generation X.

Masculinity and the Gangsta's Paradise

During the 1980s many parts of California saw significant economic growth. One of the areas disenfranchised from most of this growth was the traditional African-American hearth of the Los Angeles area. The term "South Central" dates back to days of restrictive covenants and African-American migrants being forced to live South on Central Avenue. Areas in South Central LA had not been re-built after the Watts Riots of 1965. Plagued by unemployment and a lack of opportunities, many young males in the South Central core were lured into a life of street violence and gang membership. As crime increased with a downturn in the economy after the Cold War, the pot boiled over after members of the Los Angeles police department were exonerated in the case of physical abuse of Rodney King by a jury in Simi Valley. This led to the civil uprising of 1992 where much of South Central LA was burned and looted in response to the frustration many felt in the area as they had little access to economic growth or justice.

Gangsta rap had been a threatening derivative of rap and hip hop that found some success during the 1980s. The lyrics focused on masculinity, misogyny, and the gang or "thug" life (Boyd 1996). As the civil uprising was playing out on television screens shot by news helicopters hovering over South Central, notable rappers such as Ice-T and Ice Cube were spokespersons for the plight of life in the ghetto. The civil uprising in Los Angeles brought a visceral or authentic nature to gangsta rap and provided the means to propel some of the artists into the mainstream. One of the more successful of the rappers was Dr. Dre who had been a member of one of the earliest Los Angeles-based groups NWA (Niggaz With Attitude). Dre proudly promoted Compton, California (Figure 7.1) as his home and freely spoke of gunfights, rape, police brutality, and gang culture. It was a stark contrast to much of the other music that accentuated lifestyles in California. In 1992, he joined forces with Snoop Doggy Dogg, from Long Beach, who proved to be a powerful duo in making hit records. "Nuthin' but a G Thang" (like songs of the surf era) brought a whole new lingo to the forefront with terms like "ho" (derogatory word for female), "loc'ed" (Spanglish term for crazy), "chronic" (marijuana), and "toke" (inhaling marijuana). The song stresses three important themes: gang banging or violence, having very casual sex, and bragging about the success of themselves as rappers. This was a critically important breakthrough song for gangsta rap as it peaked at #2 on the Billboard Charts despite its controversial language and adult themes.

"Wit Dre Day" was a follow-up song to "Nuthin' but a G Thang" and found almost as much popularity as the breakthrough hit. Here the artists openly criticize and antagonize their enemies who are explained as "bitches and hos," "the police," and "Luke" (Luther Campbell of the Mimi-based group 2 Live Crew). Another important location code in this song is the term "187" which is a gang term for murder borrowed from the California Penal Code for homicide. The brashness and masculinity are overwhelming in much of gangsta rap. Tupac Shakur, widely thought of as one of the more insightful rappers of this era, had a significant hit with "California Love" which featured Dr. Dre. The rap begins with an allusion to the Wild West in comparing South Central and gang justice to a previous era of mob justice over police power. It also celebrates the victory of West Coast rap over other styles from New York and Miami. "Shout-outs" (salutations) are made to other important African-American areas in California such as Oakland, Sacramento, and Pasadena which are part of this West Coast success story. While the era of gangsta rap was not exceedingly long, the images of violence and masculinity left an indelible imprint to popular notions of place in California.

Conclusion

Representations of California in popular music lyrics during the second half of the twentieth century changed quite radically from a distant, tropical, romantic paradise to a crime-ridden, drug-abusing misogynistic gang territory. This change, while stark, happened gradually and reflected political, social, and economic realities of life in some parts of the state. Substance abuse was a prevalent theme underscoring much of the San Francisco scene's music and not surprisingly, disappointment and disillusionment about the place of California emanated from music of the late 1960s similar to side effects of recreational drugs. During the 1970s, American society dealt with significant structural issues. In many ways, these economic and cultural realities hit California especially hard as the luster wore off the "promised land" or the Golden State. Much of the positive, upbeat, and celebratory imagery in the lyrics of early popular music about California possibly set up place imagery that no terrestrial landscape could "live up" to. What makes these representational changes so significant is that California's popular notions of having a mild climate and laid-back lifestyle have been in some ways supplanted by popular news commentary of drug abuse, illegal immigration and racial/ethnic hostility, and excessive cost of living. While these changes in popular image are not causing large-scale depopulation movements, they do tend to justify the folklore claim that California is the "land of fruit and nuts," meaning that many who reside there can be viewed through pop music lyrics as eccentric and odd.

The link between landscape, lyrics, and perception is critically important in representing place in popular culture. While this study takes a semiotic approach to examining Top 20 songs because of their wide airplay and high volume of sales, this is only one of many different approaches to studying geographical aspects of place in music and lyrics. Considering the banal nature of most popular music lyrics and themes, it is remarkable to reflect on how many songs (27) about California attained a Top 20 position in a range of 44 years. The multitude of songs about the state allow for listeners to craft a mental image or map of California based on their perceptions of popular music lyrics. There is something intrinsically captivating about California to the geographical imagination whether it is encased in a mindset of the Promised Land, hippies and dissention, or gang culture. Song writers and music marketers have embraced this in a lucrative yet revealing way and through popularity, the public has responded to this image creation. The combination of the interpretation of lyrics associated with emotional engagement with sounds and melodies allow for a deep, intuitive connection between listener and artist. These connections through popular culture are often a reflection of real-world experiences mediated by complex political/social relations taking place during various eras. Through analyzing the meanings embedded in music stemming from different times, we are able to excavate diverse representations of place and gain a glimpse into the popular consciousness and cultural geographic awareness of place through time.

References

Arkell, T. (1991), "Geography on Record: Origins and Diffusion of the Blues," *Geographical Magazine* 63:1, 30-34.

Bowen, D. (1997), "Lookin' for Margaritaville: Place and Imagination in Jimmy Buffett's Songs," *Journal of Cultural Geography* 16:2, 96-104.

Boyd, T. (1996), "A Small Introduction to the 'G' Funk Era: Gangsta Rap and Black Masculinity in Contemporary Los Angeles," in Dear et al. (eds.).

Carney, G. (1974), "Bluegrass Grows All Around: The Spatial Dimensions of a Country Music Style," *Journal of Geography* 73:4, 34-55.

—— (1979), "T for Texas, T for Tennessee: The Origins of American Country Music Notables," *Journal of Geography* 78:4, 218-25.

Clayton, M. et al. (eds.) (2003), *The Cultural Study of Music: A Critical Introduction* (New York: Routledge).

Dear, M. et al. (eds.) (1996), *Rethinking Los Angeles* (Thousand Oaks: Sage).

Faragher, J. et al. (1994), *Out of Many: A History of the American People, Vol. 2* (Englewood Cliffs, NJ: Prentice Hall).

Ford, L. (1971), "Geographical Factors in the Origin, Evolution, and Diffusion of Rock and Roll Music," *Journal of Geography* 70:8, 455-64.

Green, J. (1995), *The Green Book of Songs by Subject: The Thematic Guide to Popular Music* (Nashville: Professional Desk References).

Horseley, A. (1977), "The Spatial Impact of White Gospel Quartets in the United States," *John Edwards Memorial Foundation Quarterly* 15:1, 91-8.

Kong, L. (1995), "Popular Music in Geographical Analyses," *Progress in Human Geography* 19:2, 183-98.

Kramer, L. (2003), "Subjectivity Rampant! Music, Hermeneutics, and History," in Clayton et al. (eds.).

Krippendorf, K. (1980), *Content Analysis: An Introduction to its Methodology* (Beverly Hills: Sage).

Kuhken, R. and Sexton, R. (1991), "The Geography of Zydeco Music," *Journal of Cultural Geography* 13:1, 27-38.

Leyshon, A. et al. (eds.) (1998), *The Place of Music* (New York: Guilford Press).

Robinson, J. (ed.) (1997), *Music and Meaning* (Ithaca, NY: Cornell University Press).

Rycroft, S. (1998), "Global Undergrounds: The Cultural Politics of Sound and Light in Los Angeles, 1965-1975," in Leyshon, et al. (eds.).

Saunders, R. (1993), "'Kickin' Some Knowledge: Rap and the Construction of Identity in the African-American Ghetto," *The Arizona Anthropologist* 10:1, 21-40.

Shepherd, J. (1991), *Music as Social Text* (Oxford: Polity Press).

Walton, K. (1997), "Listening with Imagination: Is Music Representational?," in Robinson (ed.).

Whitburn, J. (2000), *The Billboard Book of Top 40 Hits, 7th Edition* (New York; Billboard Publications).

Wicke, P. (1995), *Rock Music: Culture, Aesthetics, and Sociology* (Cambridge: Cambridge University Press).

Part IV
Place in Music/Music in Place

The following three chapters investigate the relationship between specific artists or forms of music and the places from which they originate. In "Musical Cartographies: *Los Ritmos de los Barrios de la Habana*" John Finn and Chris Lukinbeal aim both to advance theoretical perspectives as well as produce a case study of Havana, Cuba, where they delineate four distinct neighborhoods in the Cuban capital relying on personal experiences of soundscapes. Theoretically, Finn and Lukinbeal are influenced by French philosophy, specifically Michel de Certeau's analysis of everyday life (1988). It is especially the taken-for-granted spaces in our immediate surroundings and how individuals negotiate their movements in urban environments that inspire Finn and Lukinbeal. Therefore, they distinguish between two ways of representing space: the "tour" in the de Certeauian sense where the experience of the individual is placed at the forefront as he or she travels (on foot in this case) along streets allowing for a distinct form of spatial narration that is different from how geographers usually portray the world with maps. The tour is characterized by mobility and movement, path-dependency, and subjectivity. The map, on the other hand, is (purportedly) objective, static, and non-temporal.

While the authors do not fully break with traditional geography and mapping in a Cartesian manner, the emphasis is nevertheless on the experiential, emotive, and subjective. Moreover, Finn and Lukinbeal are interested in the performative aspects of music, rather than textual analysis (which is represented elsewhere in this book, including the chapters by Daynes, Romig, and Kuhlke), and in that sense they connect with a body of literature in music geography that stresses the affect of sound (Anderson, Morton, and Revill 2005) and the interface between the performance of music and how it is received by an audience. This is where places and spaces acquire meaning. The studies that Finn and Lukinbeal build on primarily include Susan Smith's work (1997; 2000) where audible experiences are as important as visual experiences. With the sensibilities of musicians, they add an extra dimension to this body of research by equating the process of musical notation with the process of map making.

Finn and Lukinbeal's Havana contains many musical styles, but much like Billy Bragg's music was analyzed from a distinctly spatial perspective in Chapter 3, the following two chapters focus on specific artists and the places they are intimately tied to: the Red Hot Chili Peppers and Los Angeles, and the Rheostatics and Canada.

As Michael Pesses shows in "The City She Loves Me: The Los Angeles of the Red Hot Chili Peppers," the Red Hot Chili Peppers, a band blending funk, rock, punk, jazz, rap, and pop, is directly linked to the city of Los Angeles. Both city and band are inherently postmodern, according to Pesses: the band with its pastiche of both high art and the vernacular, the city with its fragmented urban form.

The Red Hot Chili Peppers weave local identities into their artwork, videos, and music. Analyzing the sounds, images, and texts of songs and videos by the Red Hot Chili Peppers, the author arrives at the conclusion that the band does not offer a straightforward critique of the city. Rather than limiting themselves to one location in the sprawling city and by addressing multiple themes, the Red Hot Chili Peppers are able to produce versions of L.A. that speaks to multiple classes and identities. It makes sense that it is Los Angeles that would produce such a postmodern rock band, and, at the same time, the sonic choices and images of the Red Hot Chili Peppers have reinforced the image of Los Angeles as a postmodern city. The two, taken together, produce a dialectic of place.

There is, of course, an overlapping geography between Kevin Romig's California from Chapter 7 and Pesses's Los Angeles. However, Pesses deals with the musically and representationally category-defying Red Hot Chili Peppers. Thus, he does not aim for an image consensus as in Romig's California, but rather builds on the conflictual messages that emanate from the Red Hot Chili Peppers. Together with the California images explained by Romig, the Red Hot Chili Peppers make up the multi-faceted collage that is postmodern Los Angeles.

In "The Geography of 'Canadian Shield Rock': Locality, Nationality, and Place Imagery in the Music of the Rheostatics" Olaf Kuhlke is implicitly arguing that Canadian national identity can more or less be distilled out of one band and its musical production. The Rheostatics offers a mirror image of Canadian society and its music can be analyzed and interpreted as a coherent and cohesive vision of Canadian identity.

The reader may ask, who is this band? A search on emusic, an online purveyor of obscure music yielded no traces of the Rheostatics. What about the search engine at rollingstone.com? The site responded "no hits" to a Rheostatics query, probably with no sense of irony. Clearly, some music has a harder time to find an international audience than others. In this case, the topical content that is made by and for Canadian ears and minds does not have a great level of supranational transferability. In a sense, this reinforces one of the points made in Kuhlke's article: Canadian separateness is reflected in the career of the Rheostatics—they are almost as iconic as Neil Young, the Band, and Leonard Cohen in Canada, but largely unknown south of the border (and elsewhere).

Some of the most important elements of "Canadianness" that Kuhlke identifies in the Rheostatics music are multiculturalism, identity through the "othering" of the United States, a political ideology (social democracy "light") that emphasizes the provision of social services by the state, and the use of northern landscape and nature imagery as "glue" for a fragmented nation. The Rheostatics offer a progressive vision of Canada, an alternative discourse in times of neoliberal hegemony, and more complex use of landscape metaphors compared to the musicians in ambient Australia who painted aural pictures of another place where Anglo-Saxons met the indigenous. For the Rheostatics, Canadianness reigns supreme, and there is little room for separatism. In fact, the cohesiveness of Canada the Rheostatics presents is perhaps imagined. The Rheostatics are reflective of the Canadian center—they hail from Ontario—but the contrast could not be greater compared to the ideology present in Sara Beth Keough's following chapter (Chapter 11) on Newfoundland radio. The musical image of Canada is suddenly reshaped when viewed from the Atlantic periphery.

References

Anderson, B., Morton, F., and Revill, G. (2005), "Practices of Music and Sound," *Social & Cultural Geography* 6:5, 639-44.

de Certeau, M. (1988), *The Practice of Everyday Life* (Berkeley: University of California Press).

Smith, S.J. (1997), "Beyond Geography's Visible Worlds: A Cultural Politics of Music," *Progress in Human Geography* 21, 502-29.

Smith, S.J. (2000), "Performing the (Sound)world," *Environment and Planning D: Society and Space* 18, 615-37.

Chapter 8

Musical Cartographies: *Los Ritmos de los Barrios de la Habana*

John Finn and Chris Lukinbeal

Approached from the outside, Cuba's shared history and culture unifies the island and gives it its *Cubanía*, or Cubanness. In the twenty-first century, *Cuban* describes a people, a language, a culture, an identity. Approached from within Havana's neighborhoods, though, the view is quite different. As the scale of inquiry shifts to the city blocks, sites, sounds and smells rise to prominence and grand cultural generalizations seem irrelevant as previously invisible distinctions emerge. In Havana, these distinctions are clearly seen, heard, and felt in the rhythms of the neighborhoods. Musical definitions of Havana are at their richest and most personal at the neighborhood scale where shared cultural histories create unique soundscapes. At this scale the streets have rhythm as bodies, instruments, cars and buildings provide the structures that carry the beat.

Geographers have long used mapping as a tool to reveal the geographical nature of music. This is most clearly seen in work on the diffusion of styles (Jackson 1952), the locations of music-related phenomena, such a associations, competitions, and actual musicians (Crowley 1987), and geographic attributes of hierarchies of styles and sub-styles (Burman-Hall 1975). These and many other studies in music geography have built a strong base from which current research moves forward. Indeed, it is because of these foundational works that music is now considered an important realm of geographical study. This has meant an impressive increase in academic literature in the area, including multiple sub-disciplinary reviews (e.g. Kong 1995; Smith 1997; Hudson 2006), various special issues (*Transactions of the Institute of British Geographers* 1995; *Social & Cultural Geography* 2005; *GeoJournal* 2006), and multiple volumes dedicated to the geographical analysis of music (Stokes 1997; Leyshon et al. 1998; Perrone and Dunn 2002).

Rather than jettisoning traditional approaches to music geography, the purpose of this chapter is to reinvigorate it by exploring the neighborhoods of Havana through music's metaphoric relation to maps and tours. The map and the tour are two distinct approaches to spatial narration (de Certeau 1988). De Certeau's (1988) two modes of spatial narratives draws from the linguistic research of Linde and Labov (1975) which presents ways that people understand and describe their surroundings. The map is seeing from above, the tour is moving within. Through mapping, space is presented in an objective manner, whereas the tour incorporates subjective matter. To Linde and Labov (1975), the map presents familiar areas narrated from above, representing location through an abstract model that relies on the god-eye trick (Haraway 1996). The god-eye trick refers to the orthographic view of the world presented through cartographic representation where the viewer is situated everywhere and nowhere at the same time. For instance, a New York City resident in Linde and Labov's (1975, 929) study described his living space as "a huge square with two lines drawn through the center to make like four smaller squares ... In the two boxes facing out in the street you have the living room and a bedroom ..." This and other maps model reality in an abstract geometric form. Mapping requires objective description and the belief in mimesis; the tour requires movement through space, where subjectivity, experience, emotions, knowledge, and valuation of the traveler plays a central role. The "map," along with descriptive fieldwork, typifies traditional cultural geography tools to approach music. In contrast, the "tour" incorporates the performative, affective role of music in place-making.

In what follows we show how the map remains a powerful tool in the analysis of regional music geographies. We then turn our attention to the tour to show how it can reveal the performative and experiential aspects of music in place-making. Finally, we apply these two spatial narrative practices to a musical cartography of Havana in an effort to show that together, the map and the tour can take us beyond the primacy of the visual in studying phenomena that are first and foremost auditory.

Musical cartographies

The idea of applying a cartographic metaphor to music may seem problematic in that the former is predominately a visual representation and the later is an auditory–artistic expression. However, we draw our inspiration for such a comparison from the recent book, *Cartographic Cinema*, by Tom Conley (2007) who suggests that cartography can provide a new way of seeing and understanding cinema. We do so because cinema is an audio-visual representation and artistic expression. While cartography and music/cinema are representational practices that are quite foreign to one another, they are nonetheless, all involved with "locational imaging" (Conley 2007; Buisseret 2003). Locational imaging refers to the representational practices of situating the viewing subject within the places that they represent. This practice is mediated by the production, style, and aesthetics of authorship and the perception, cognition, and affect on reception. While cartography relies on vision and cinema blends sight and sound, music is primarily auditory. In narrative cinema, music's role is to animate place and provide an emotional context for how places should be experienced. It could be argued that music is involved with locational imaging in that it requires authorship and reception as well as provides information about place. But, without the reliance on the visual, the metaphoric relationship between maps and music seemingly breaks down. This is because maps seek to de-animate space through the god-eye trick, to remove identity, subjectivity, and narrative from its representation to create an objectified object. Yet sensory perception cannot be objectified or wholly removed from cinema and music because they are practices that require embodied participation. A musical cartography must engage both real and mental maps of sites, memories, experiences, preferences, observations, facts, fictions, feelings, and emotions. Furthermore, musical cartographies must also blend active performances with static representations.

While it is often the performative, experiential qualities of music that first come to mind, music is most often modeled through musical notation. Notation is the musician's road map to any piece of music. This notation is literally a model of the reality that a song will become in performance. Thus like road maps, musical notation can be followed strictly: in the same way that a city bus moves through space and time on a particular route and schedule, a symphony performs Mahler's Symphony No. 5 with tempos and dynamics prescribed in a different space and time. But while musical notation provides the musician's route, this notation is only—and can only be—the mathematics of the music. In supposing to be objective models of reality, maps and musical notation both occlude the processes that gave rise to their respective existences. Maps seemingly display objective spatial information, such as absolute locations, distances measured along lines, or the extent of polygons, so much that the process of information selection, compilation, and manipulation is rendered invisible. Ingold (1993, 154-5) articulates the way experience is lost through mapping by describing a surveyor who:

> experiences the landscape much as does everyone else whose business of life lies there ... The distance between two places, A and B, is experienced as a journey made, a bodily movement from one place to the other, and the gradually changing vistas along the route. The surveyor's job, however, is to take instrumental measurements from a considerable number of places, and to combine these data to produce a single picture which is *independent* of any point of observing [original emphasis].

According to de Certeau (1988), maps colonize space, thus removing the performative aspects of a tour. We add that in music, the relationship is much the same: the musical notation that represents the final absolute version of a song is largely divorced from the creative and performative processes that went into its composition, or that will go into its future performances.

The occlusion of the productive process, however, has different connotations for musical notation and mapping. On one hand, the cartographer is condemned to anonymity by projecting both power and agency onto the map and map user. On the other hand, musical notation can immortalize a composer through performance capturing her works for future generations. In occluding the producer of the map/musical notation, these practices not only delineate the terrain of analysis, but also simultaneously limit engagement with that terrain. Both the map and musical notation are forced to conform to their own standards of practice and mathematics, independent of the object that they are designed to model.

Understanding music strictly through notation, the actual performance of the music—the act of hearing and listening to organized sound—becomes secondary to the musical text. S.J. Smith (2000, 618) observes that musical performances are often attempts "to realize the written text exactly as the composer intended it to sound. The authenticity of the performance is thus judged by the accuracy with which the written notes, and the markings that go with them, are conveyed." By focusing specifically on music-related texts (most obviously, the musical notation, but also song lyrics and other textual aspects of music), analysis of musical performance and its effects on those who hear it are completely lost. Adorno (1993) suggested that unlike language, music is interpreted in its performance. To B.R. Smith (1999, 112), this meant that music is what is heard, "not what is printed on the page." This idea is perhaps most clearly presented in Grossberg's (1984) ruminations on rock and roll where he decouples textual and semantic analysis of songs and the affective results on listens. To him, the effect of both recorded and live rock and roll cannot be measured through text: "It is not that rock and roll does not produce and manipulate meaning but rather that meaning itself functions in rock and roll affectively, that is, to produce and organize desires and pleasures" (Grossberg 1984, 233). In geographical analysis of music, what the music *does* should be at least as important as what the musical text says.

Seeking to "map" music limits it to textual analysis and descriptive geography. In doing so, the creative, performative, and interaction aspects of music are secondary to hermeneutics or description. Mapping forces the terrain of music to be a material or a non-material cultural product which can be modeled in an abstract Cartesian space. Based on this traditional approach to music geography, a map of Havana's musical geographies would only reveal cultural and social practices that have an absolute location (see Figure 8.1). For example, the location of the EGREM recording studios where the *Buena Vista Social Club* recorded their 1996 Grammy-award winning tribute to traditional Cuban music, and also where North American greats such as Nat King Cole made historic recordings in the 1940s-1950s, represents one node in a musical topologic network embedded in Havana. The birthplaces of legendary Cuban musicians such as Beny Moré, Compay Segundo, and Chucho Valdés, and the sites of their musical performances would provide further point data. Whereas a musical cartography of New York City would be incomplete without the Birdland and Carnegie Hall, the locations of the *Teatro Amadeo Roldán*, the *Gran Teatro*, and the *Jazz Cafe* are fundamental to Havana's musical map. The site of the Sunday morning rumba by the *Conjunto Folklórico Nacional*, and the famous *Callejón de Hamel* (see Figure 8.2) are essential to the accurate representation of Havana's vibrant Afro-Cuban scene. Even the diffusion of the African influence in traditional Cuban style *danzón* could be depicted with a qualitative flow map. African and French influence in Haiti eventually impacted eastern Cuba during the slave rebellion at the end of the eighteenth century before diffusing across the island, adapting a Cuban feel, and becoming Cuba's national music by the early twentieth century. In short, there is an unlimited amount of spatial data to draw from to map Havana's musical

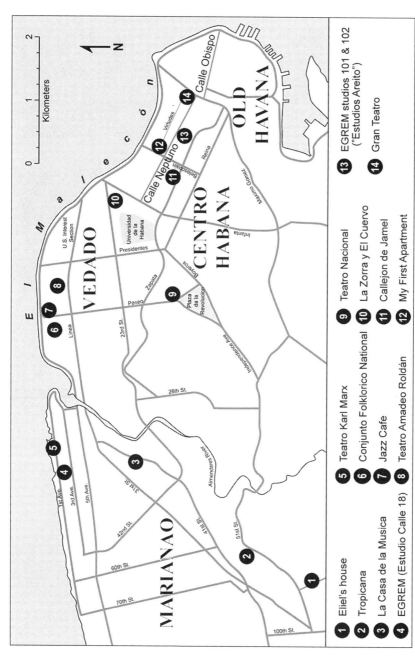

Figure 8.1   A musical map of Havana

Figure 8.2   Rumba dancers at *Callejón de Hamel, Centro Habana*
             © Megan Resch

geographies. It would, however, require that a cartographer sift through the data to determine how classification and generalizations should be made.

The cartographic metaphor does have its limits when applied to music. S.J. Smith's (1997, 504) critique on writing about music seems most appropriate:

> Writing about music is like dancing about architecture, listening to a ballet or feeling the texture of a painting – it might be helpful, but it is not the best, most direct or most appropriate way of illustrating the power of the art ... Music conveys meaning through rhythm, melody and harmony ... Music has to be heard to be understood.

Since music is first and foremost aural, writing about music, or mapping music, appears to appropriate one sensory representation with another. That being said, years of tradition in cultural geography and landscape studies have privileged visuality over other sensory ways of knowing, greatly diminishing the effect of the non-visual. To S.J. Smith (1997, 503), this should not come as a shock in a discipline that is a "quintessentially visual enterprise, traditionally using [visual] observation as the route to knowledge, and regarding sight as the measure of truth." The axiom of needing to see in order to believe is a clear product of the privileged position of visuality. A map of music's visual attributes cannot capture rhythm, melody, and harmony—the specific mediums of communication that conveys meaning in music. For that we turn to the tour.

The tour

While geographers frequently model space in cartographic form, when prompted to describe their living spaces, only three percent of participants in Linde and Labov's (1975) aforementioned study used this approach. The other 97 percent of participants sought to produce a virtual "tour," of their living space:

> A tour is a speech act *which provides a minimal set of paths by which each room could be entered* ... The imaginary tour does not always take the listener into each room. In [one example] the listener walks in the front door and is in the hallway. He comes to the first door, but he doesn't enter it. He passes the kitchen and the bathroom, sees how to enter them but does not do so, and finishes in the living room ... The task of the speaker, then, is to traverse for this imaginary visitor the spatial network formed by these vectors and rooms (Linde and Labov 1975, 930, original emphasis).

The tour actively seeks to provide the visitor (i.e. the reader) with the experience of being in the particular place, of walking through, peering into rooms and negotiating hallways and tight spaces.

De Certeau (1988, 119, original emphasis) asserted that in spatial narratives, "description oscillates between the terms of an alternative: either *seeing* ... or *going.*" This is also true in music as musicians oscillate between reading and improvising, seeing and going, the map and the tour. Indeed, not all music can be conceived through musical notation. Jazz, for instance, is a form based largely on improvisation. While the basic melody may (or many times may not) be scribbled on a sheet of paper, the song is largely created through performance. That is the magic of improvization in live music—it is invented on the spot and experienced differently by every single person present at its creation. Miles Davis provided his musicians with only "sketches" (Davis and Jones 1990, 234) of musical notation when recording the classic album *Kind of Blue* (Davis 1997 [1959]). The music that was recorded was largely the product of the tour, as Davis, John Coltrane, Bill Evans, and others moved through sound, improvisationally experimenting with colors and rhythms, peering down sonic hallways, and guiding the listener through their musical subjectivities.

According to de Certeau (1988), the tour is walking, doing, and moving; it is fully experiencing the performance of the musicalized space. Where the map is the representation of the finished product of a cartographer/composer, the tour is the corporeal experience of a place. Knowledge and experience of a place drives the subjective experience just as much as the affective nature of the interaction with music in place.

In the next section, we oscillate between objective mappable descriptions and a subjective tour of Havana's neighborhoods. To do so, we draw from John Finn's personal experiences, mental map and knowledge of the cultural geography of Havana. We invite you to stroll with us down the streets of rhythm.

Figure 8.3   Colonial Spanish architecture, *Centro Habana*
                © **Megan Resch**

A guided map and tour of Havana

This section of the chapter—the "walk" through Havana—is based on more than a year that I lived in that city. The directions that I give do not come from a single journey made, but rather are my collective movements as I lived, worked, and most importantly, played music, during my tenure there. It is important to state, however, that while the meaning I convey here comes from my own impressions and are thus subject to my personal interpretations, the cardinal directions, streets, place names, and other features that I detail in the landscape are all real.

My first apartment in Havana was near the corner of *Calle Belascoaín* and *Calle Virtudes*, in the crumbling neighborhood of *Centro Habana*. The streets of this inner-city neighborhood are lined with once elegant Spanish row houses that were, in a different era, the focal point of the Cuban bourgeoisie (see Figure 8.3). Now it is among the most overcrowded and decrepit sections of the city. Multiple families subdivide these former Spanish mansions, and often a single room is cut both horizontally and vertically (a false ceiling creates a second floor within the single room) and houses an entire family. In my first week there, one of the centuries-old mansions gave way to years of abuse, overpopulation and under-reparation, leaving a pile of debris some four meters tall, ten meters wide and 40 meters deep wedged between two buildings awaiting a similar fate.

While structures seem to collapse and disintegrate all around, a more vibrant place in the city is hard to find. This ransacked neighborhood is the heart of the

Figure 8.4   Claves in the rumba, *Callejón de Hamel, Centro Habana*

Afro-Cuban community. Going west from my apartment and left at the first light, past the remains of the collapsed building—still there six months later—and past the warehouse on the corner that has become permanent temporary home for hundreds of displaced people (still *there* six years later), is *Calle Neptuno*, the main drag through the neighborhood. When walking this street, it only takes moments to see a young person dressed completely in white as part of the initiation to S*antería*, or a priest of the religion draped in colorful beads, each combination of colors representing a different *santo*, black and red for *Eleggua*, blue and white for *Yemaya*, white and red for *Chango*. And at night this neighborhood is overtaken by Afro-Cuban celebrations, often in honor of specific saints on their saint-day. Invariably, there are drums. The *clave*, or the "key," whose simple three-two or two-three pattern is the rhythmic basis for all Afro-Cuban music, guides the ensemble as the high-pitched, metallic pulse of the *guataca*, usually made from a hoe blade, drives the music forward. The complex rhythms of the wooden *cajón*—literally a box that the percussionist sits upon while playing between his legs—weave through a musical time that is not quite square enough to be a duple beat, yet not rounded enough to be a triple feel (see Figure 8.4). The *chekeré*, a hollowed gourd strung with beads and played as a shaker completes the ensemble as the pulsating rhythms, *cantos* of singing participants, and footsteps of the dancing mass of bodies flood from interior rooms and courtyards, through open

Figure 8.5   *La Habana Vieja*, a UNESCO World Heritage Site
© Megan Resch

windows and doors, into the street. Living here for so many months gave me a new understanding of what S.J. Smith (2000, 616) refers to as the contested place of music "between the myth of silence and the threat of noise," as the soundtrack of this *barrio* always seemed to organize itself to reflect the musical and rhythmic hearth of the Afro-Cuban.

Going east toward the Havana Bay, *Calle Neptuno* eventually leads to *La Habana Vieja* (Old Havana), the original site of the Spanish city created almost 500 years ago. As *Centro Habana* fades into Old Havana, the rhythms of the streets change accordingly. The architecture and style of the buildings are even older than *Centro Habana*, and Old Havana feels like an entirely different world. This historical part of the city, a UNESCO World Heritage Site, has been the recipient of millions of dollars of investment for its restoration (see Figure 8.5). As buildings are marked for renovation, residents are moved out (they are told it is temporary) and the construction crews move in. These ancient structures are gutted and restored to their former grandeur. But more common than the return of residents has been the arrival of hundreds of thousands of tourists annually. Colonial mansions that formerly housed multiple families become hotels and restaurants and galleries to accommodate the newer breed of Old Havana residents—Canadians, Italians, Germans, and the occasional brave American.

As in *Centro Habana*, music is everywhere. But with its newly restored architecture and its quickly changing demographic, a very different sound flows through the streets of Old Havana. The new sound of Old Havana was defined

Figure 8.6 *Malecón*

by the 1996 release of the *Buena Vista Social Club*, an album (falsely) identified as belonging to Ry Cooder, and hugely popular throughout the United States and Europe. This music, the traditional Cuban *son*, is rooted in the songs of European troubadours infused with the percussion and rhythm of Cuba's African heritage. After reaching its height of national popularity in the 1940s and 1950s, this sound largely faded out of the popular Cuban conscience. But today, moving through the restaurants, lounges, and open-air cafes of Old Havana's *Calle Obispo*, or Bishop Street, the sound is distinctively *son*. This avenue, now open only to pedestrian traffic, was where the neighborhood's restoration started, and has become a must-see on every tourist's itinerary. Here old Cuban men in short-sleeve *guayabera* shirts and smoking Cuban cigars form the quartets and quintets that fit the *mojito*-sipping tourist's pre-conceived notion of what an authentic Cuban neighborhood should look like, and what real Cuban music should sound like. The sound of Old Havana is traditional Cuban culture commodified and made new, packaged for the international visitors whose dollars are the lifeblood of this quarter.

Old Havana's Bishop Street ends where one of the city's most famous landmarks, the *Malecón*, begins (see Figure 8.6). From here running westward, this eight-kilometer seawall protects the city from the high tides of the Straights of Florida and is the northern boundary of Old Havana and *Centro Habana*. Looking due north from here it feels as though the lights of Miami, and the underlying capitalism dominated by a very different *cubanía*, are just out of sight. After about three kilometers the *malecón* passes directly in front of the famed US Interest Section, the focal point of US–Cuban antagonism on the island, and eventually

Figure 8.7   Young Cuban singers performing with the popular *timba* group NG La Banda © Megan Resch

leads to the *Almendares* River. Here, two bridges span the river and two tunnels pass underneath. Passing through either of the tunnels, roads that previously paralleled the coast start moving slightly inland and uphill as they continue west. They lead first through *Miramar*, the embassy zone of the city, and finally into *Marianao*, one of the city's largest residential neighborhoods—and a neighborhood too out of the way for most tourist traffic.

For almost a year I made this journey to *Marianao*, deep in the heart of Havana, to the home of my percussion teacher, Eliel Lazo. During these bi-weekly pilgrimages I was often stopped for quick conversations with curious local residents (my blond hair, blue eyes and funny accent kept me from being mistaken for a Cuban). They wanted to know why I had strayed from the beaten tourist path (that is, between the beach and Old Havana). When I told them I studied percussion in this neighborhood, they immediately understood why I made the relatively long journey by crowded bus twice a week. *Marianao* is to popular Cuban music what Harlem was to bebop in the 1950s and '60s. In *Centro Habana*, *Bembé* and other forms of Afro-Cuban music seep from the crumbling structures, and in Old Havana the traditional *son* saturates the air. But in *Mariano*, a new style—*timba*—is the sound of the *barrio* (see Figure 8.7). This song-style incorporates instruments from North America—drums, electric guitars, and synthesizers—to the Cuban line-up of congas, timbales, horns, piano, and other percussion. Its velocity and ferocity, complex rhythms and syncopation, and its sensual, if not blatantly sexual content are especially appealing to a new generation of Cubans that have come of age in a post-Soviet Cuba, and create a sound representative of the economic and political hardships of the post-1990 Special Period. The sound of *timba* and the social conditions of the contemporary Cuban youth converge in *Marianao*. This sound emerges from houses and apartments where groups are rehearsing. It blasts

Figure 8.8   The *Vedada* neighborhood in Havana © Megan Resch

from the pieced-together stereo systems in '57 Chevies and '77 Ladas alike. This is the new Cuban sound, it is the lived experience of a new generation manifested through music, and its home is not only in the streets and houses and dancehalls of *Marianao*, but performed throughout the neighborhood by its residents.

Leaving *Marianao* and travelling back in the general direction of Old Havana and *Centro Habana*, this time crossing the 23rd Street bridge, buses, shared taxis, and the occasional private car move in unison toward downtown Havana, the most international and cosmopolitan sector of the city (see Figure 8.8). By day this area, known as *Vedado* to local residents, defies easy categorization. In addition to a large residential sector, the neighborhood is made up of urban high-rises, hotels, national and international businesses, banks, shopping centers, the national university, and countless government installations. Developed more recently than many other parts of the city, wider streets allow for heavier traffic and ample sidewalks provide abundant space for the multitude of pedestrians negotiating their way through the city. The international feel that downtown Havana possesses is a result of much more than tourism: if Old Havana is international in the sense of foreigners in Hawaiian shirts and tropical drinks taking rides in horse-drawn buggies, *Vedado* is international in the sense of foreign businessmen and diplomats in dark suits penetrating the city in oversized black sedans with tinted windows and foreign flags on the antennas. While the rest of the country may lag behind, this neighborhood is Cuba in the twenty-first century. This place is Cuba after the fall of the Soviet Union, flexing whatever international muscle it can muster.

**Figure 8.9   Pianist Robert Fonseca performing with Omar Gonzalez
              (bass), Javier Zalba (sax), and Ramsés Rodriguez (drums) in
              La Zorra el Cuervo during the 2008 Havana International Jazz
              Festival © Megan Resch**

For everything that sets downtown Havana apart from the rest of the city, this neighborhood, like every other, has a sound. On the corner of 23rd Street and O, the center point of *Vedado*, there is what seems to be a bright red British phone booth on the street corner. Upon entering, instead of a phone, a stairwell descends into a tiny basement club and restaurant—Havana's hottest spot for Cuban and international jazz. This club, *La Zorra y el Cuervo*, *is* downtown Havana by night, a musical microcosm of the emerging internationalism visible on the streets during the day. This club is closer to New York City than anything else on the island, and the sounds are not those of Cubans performing a commodity for mass tourist consumption. Rather, the sound is of Cubans performing as a part of a global music scene, on stage with their international counterparts in open jam sessions that can last well into the morning (see Figure 8.9). This is where, during the Havana International Jazz Festival (locally referred to as Jazz Plaza), international and North American jazz icons perform on stage with Cuban expatriate musicians and any number of local Havana jazzers. The downtown Havana jazz scene, as played out in *La Zorra y el Cuervo*, is the idealization of the neighborhood by day. This scene, revolving around a specific sound in Cuban jazz, has resolved the political problems that have plagued US–Cuba relations for almost five decades. It is a scene that has gotten past socio-economic differences, which, upon exiting the club, are omnipresent in the glaring disparity on the street. It is a scene that

Figure 8.10   Sunday Mass in Old Havana's Cathedral
   © **Megan Resch**

Figure 8.11 *Santero* in *Centro Habana* © **Megan Resch**

celebrates its Afro heritage instead of being mired in the racial problematic that continues to afflict the rest of the country, and indeed the entire world.

Conclusion

For more than 500 years Cuba has been the site of cultural integration. Religious syncretism has led to the practice of *Santería*, in which African gods take the names and physical forms of Catholic saints (see Figures 8.10, 8.11). Linguistic mixing has created a distinct Cuban dialect that infuses African words into the dominant Spanish vocabulary. The rhythms and melodies of diverse African cultures, ranging from present-day Senegal to the Congo River Basin, have fused with Iberian sounds and song-styles, and more recently, tremendous cultural influence from the north. The result of this cultural miscegenation is an island whose music, together with Brazilian and American popular music, "has had the greatest influence on popular music throughout the world" (Uribe 1993, 11). Accordingly, geographical approaches to music can only do justice to the actual music itself if they take into account not only analysis of visual and textual attributes of sound, but also the affective reality of its creation and performance.

Thus treating music "as something else *to be seen* diffusing in space, trickling down hierarchies, attached to the landscape" (S.J. Smith 1997, 504, original emphasis) is only one part of understanding the geographies of musicalized space. In *Centro*

Habana, the rhythm of the street—the drums and voices, the movement of feet across concrete slab floors, the almost nightly Afro-Cuban celebrations and the partying that goes along with them—are all essential elements to the (re)production of this place. Music is equally important to place-making as the non-musical elements of overcrowded conditions and crumbling homes and buildings. In Old Havana—while so physically close to *Centro Habana* that the line between neighborhoods changes depending on whom you ask—the music, the actual sound emerging from buildings, homes, restaurants, and hotels, draws a clearer line of distinction than any mapped boundary. *Son* is central in demarcating the neighborhood, and its importance in transforming this space into a place of tourism would be hard to overstate. In Old Havana it is also this sound and style that employs so many local musicians, forcing them to retrace the musical traditions of their parents and grandparents. Moving west, crossing much of the city, and deep into the heart of *Marianao*, sound and music once again cannot be separated from the character of the *barrio*. Here *timba* is the musicalized lived experience of a generation of Cubans in a post-Soviet era. Finally, it was not through coincidence or simple convenience that this "walk" ended at a tiny basement music club in *Vedado*. More than anywhere else in the city, it is here that sound transforms space into place. Sound breaks through politics, bends perceptions of international affairs and geopolitics, and moves past racial and social problems. This power of sound is central to understanding the subjectivities of performance, the actual sound of a place.

Oscillating between the concepts in spatial narration of "map" and "tour" enables us to use foundational concepts and approaches in music geography in combination with the performative and affective attributes of actual sound so often left out of academic view. Putting the map and tour in dialectical tension is one way to approach a musical cartography that goes beyond music's visual attributes. Through simultaneously seeing *and* going we start to feel *los ritmos de los barrios de la Habana*.

References

Adorno, T.W. (1993), "Music, Language, and Composition," *The Musical Quarterly* 77, 401-14.

Buisseret, D. (2003), *The Mapmaker's Quest: Depicting New Worlds in Renaissance Europe* (New York: Oxford).

Burman-Hall, L.C. (1975), "Southern American Folk Fiddle Styles," *Ethnomusicology* 19:1, 47-65.

Carney, G.O. (ed.) (1987), *The Sounds of People and Place: Readings in the Geography of American Folk and Popular Music* (Lanham, MD: University Press of America).

Conley, T. (2007), *Cartographic Cinema* (Minneapolis: University of Minnesota Press).

Crowley, J.M. (1987), "Old-time Fiddling in Big Sky Country," in G.O. Carney (ed.).

Davis, M. and Jones, Q. (1990), *Miles: The Autobiography* (New York: Touchstone).

Davis, M. (1997 [1959]), *Kind of Blue* (Sony: B000002ADT).

de Certeau, M. (1988), *The Practice of Everyday Life* (Berkeley: University of California Press).

GeoJournal (2006), "Special Issue on Geography and Music," *GeoJournal*: 65:1/2, 1-135.

Grossberg, L. (1984), "Another Boring Day in Paradise: Rock and Roll and the Empowerment of Everyday Life," *Popular Music* 4, 225-58.

Haraway, D. (1996), *Simians, Cyborgs and Women: The Reinvention of Nature* (London: Free Association Books).

Hudson, R. (2006), "Regions and Place: Music, Identity and Place," *Progress in Human Geography* 30:5, 626-34.

Ingold, T. (1993), "The Temporality of Landscape," *World Archaeology* 25:2, 152-74.

Jackson, G.P. (1952), "Some Factors in the Diffusion of American Religious Folksongs," *Journal of American Folklore* 65:258, 365-69.

Kong, L. (1995), "Popular Music in Geographical Analysis," *Progress in Human Geography* 19:2, 183-98.

Leyshon, A., Matless, D. and Revill, G. (1998), *The Place of Music* (New York: Guilford Press).

Linde, C. and Labov, W. (1975), "Spatial Networks as a Site for the Study of Language and Thought," *Language* 51, 924-39.

Perrone, C. and Dunn, C. (eds.) (2002), *Brazilian Popular Music and Globalization* (New York: Routledge).

Social & Cultural Geography (2005), "Special Issue on Music Geography," *Social & Cultural Geography* 6, 639-766.

Smith, B.R. (1999), *The Acoustic World of Early Modern England* (Chicago: University of Chicago Press).

Smith, S.J. (1997), "Beyond Geography's Visible Worlds: A Cultural Politics of Music," *Progress in Human Geography* 21, 502-29.

—— (2000), "Performing the (Sound)world," *Environment and Planning D: Society and Space* 18, 615-37.

Stokes, M. (ed.) (1997), *Ethnicity, Identity and Music: The Musical Construction of Place* (Oxford: Berg).

Transactions of the Institute of British Geographers (1995), "Themed Issue: The Place of Music," *Transactions of the Institute of British Geographers* 20, 423-85.

Uribe, E. (1993), *The Essence of Brazilian Percussion and Drum Set* (Miami: CPP/Belwin).

Chapter 9

The City She Loves Me: The Los Angeles of the Red Hot Chili Peppers

Michael W. Pesses

Let me play with your emotions / For nothing is good unless you play with it Yeah / Fly on / Fly on, baby ("Mommy, What's a Funkadelic," Funkadelic 1970)

Introduction

In the 1980s, the Los Angeles music scene had its share of "hair bands," but it also had a growing number of bands that resisted "glam" and began something new. *Jane's Addiction, Fishbone, fIREHOSE*, and the *Red Hot Chili Peppers* all borrowed from rock n' roll, punk rock, funk, country, jazz, and more to help launch an "alternative" sound. Their work differed from Seattle's "grunge," but both sounds contributed to a new movement in popular music (see Bell 1998 for a discussion of alternative music and the Seattle contribution). Gill states "the various trends and styles of rock derive initially from conditions specific to particular regions, with local dance and bar bands often being the principal sources of innovation and change" (1995, 17). This is the case with 1980s Los Angeles, but I would argue that the Red Hot Chili Peppers were not only a part of a regional sound, but that the place itself became a central presence in their music.

The Red Hot Chili Peppers have never hidden their relationship with Los Angeles. Bassist Michael "Flea" Balzary puts it bluntly in the band's 1999 VH1 *Behind the Music* episode, "We're L.A. to the bone." Throughout over twenty years of existence, the band has woven the physical and cultural geography of L.A. into their songs, videos, and mythos. Be it an irreverent party anthem or biting social commentary, a Red Hot Chili Pepper's song relies on Los Angeles to create context. The band uses postmodern techniques of playfulness, pastiche (that is, a compilation of influences and ideas), and acknowledges coexisting and/or competing realities. In doing so, the Red Hot Chili Peppers have produced a unique Los Angeles. At the same time, the postmodern structure (Dear and Flusty 1998) and essence (Soja 1996a) of Los Angeles (e.g. resisting traditional urban form) has, in turn, informed their music. The two are in constant dialectical tension and the result for the band has been an inherently geographic artistry.

Music, literature, and film geographies

A reoccurring theme in the study of music from a geographic perspective is that not enough has been done to advance this perspective, or that it needs to advance in a new direction (Kong 1995, Leyshon et al. 1995, Zelinsky 1999, and recently Hudson 2006). Popular American music still seems to be an untapped source of geographic analysis. While there is a growing amount of literature on postcolonial "world music" and the use of hybridized music for tourist consumption (Connell and Gibson 2004; Gibson and Dunbar-Hall 2000; Greene 2001), human geography has paid little attention to popular American music since the 1970s (Ford 1971; Gill 1995). Some work exists on the genesis and diffusion of particular sounds in American music (Bell 1998; Gill 1995; Graves 1999; Stump 1998). Evoking place through popular music is of interest to geographers, as is the thematic analysis of lyrics through a geographic lens (Kong 1995). Very little has been written on the role place plays as an *influence*, and even as a protagonist, not simply a setting, in American popular music (however, see Moss 1992; Forman 2000). This is surprising as "landscape ... provides a context, a stage, within and upon which humans continue to work, and it provides the boundaries, quite complexly, within which people remake themselves" (Mitchell 2000, 102). Furthermore, landscape is "an ongoing *relationship* between people and place" (Mitchell 2000, 102, emphasis in original).

A few exceptions are Ford and Henderson (1978) who explored how images of a place were shaped by music:

> The perceived qualities of places and types of places may change with their changing appropriateness for popular music themes. The hypothesis is made that songs both reflect and influence the images people have of places and that these songs have changed people's attitudes significantly toward places is one important step toward understanding the "geography of the mind" (Ford and Henderson 1978, 292).

Gumprecht also studied the relationship between place and the musician in popular American music in his analysis of the West Texas landscape's influence on Joe Ely, Butch Hancock, and Terry Allen. "Geographers who want to better understand the cultural landscape and how humans react to it ... would be wise not only to *look*, to attempt to *read* that landscape, but to *listen*, to try to *hear* what sound, including music, can tell us about place" (Gumprecht 1998, 78, original emphasis). Related but different work addresses the production of musical *spaces* or "soundscapes" (Aitken and Craine 2002; Valentine 1995). The sounds we hear out of our car stereo, over a supermarket loudspeaker, or at a concert can alter our perceptions of the material landscape. These musical 'scapes have been explored in terms of affect and emotion, but specific and fixed places are rarely discussed (admittedly so in Aitken and Craine 2002) in such studies. This chapter will attempt to explore both

mental perceptions of the Red Hot Chili Peppers' music as well as its references
to concrete places.

Just as popular American music is rarely explored in geographic terms, when it
comes to using sensory consumption and production of place in human geography
in general, the sonic takes a backseat to the visual (Valentine 1995). Sight plays a
key role in how we interpret our world, yet without sound, "space itself contracts,
for our experience of space is greatly extended by the auditory sense which
provides information of the world beyond the visual field" (Tuan 1974, 9). As I
write this, I hear a bird singing outside my office, though I cannot see it. I hear
cars in the distance, the rumble of the central heating unit, the ticking of the clock.
Without sound, my world is much smaller. I look at the computer screen, but I hear
lyrics about crashing surf (from the song "Road Trippin'," *Californication* 1999)
or untouchable basketball skill (the song "Magic Johnson," *Mother's Milk* 1989).
These lyrics and the echoing electric guitar, underlying funk bass, and almost
tribal drums extend my space from my desk located in the San Joaquin Valley to
the Los Angeles of the Red Hot Chili Peppers.

As visual elements are often the focus of media geographies, it is useful to
explore studies of literature and place and film and place in an attempt to transfer
these theories to the music of the Red Hot Chili Peppers and Los Angeles. For
example, Tuan speaks of the permanence and power of the literary text as producing
images of place. London's literary history has managed to evoke both real and
unreal feelings in tourists (1991, 690-91). Is Los Angeles much different? While
literature may not always spring to mind (although fans of noir writers Raymond
Chandler, Dashiell Hammett, and James Ellroy may disagree), Los Angeles has
appeared in countless forms of popular culture. The quintessential L.A. diner plays
a major role in the bookend scenes of Quentin Tarantino's *Pulp Fiction*. Steve
Martin's *L.A. Story* shows the surreal and absurd sides of the city (e.g. driving to
your mailbox). Popular music also utilizes Los Angeles as setting, from the Beach
Boys' coastline to N.W.A's inner city. As these examples reveal, the same city can
be envisioned as two completely different places based on not only the lyrics, but
also the sonic presentations. "Straight Outta Compton" would not have had the
same effect without the layering of Dr. Dre and Easy-E harmonizing behind Ice
Cube's lead vocals.[1]

It is worth mentioning at this point that the Red Hot Chili Peppers are well
known for their nonsensical lyrics. When Anthony Kiedis raps, "Oh ... ticky
ticky tickita tic tac toe / I know ... everybody's Eskimo" ("Warlocks," *Stadium*

---

1   For those not familiar with the music referenced, in the 1960s, the Beach Boys used
lyrics about Southern California's young beach culture (surfing, fast cars, and attractive
girls) and tightly layered harmonies from the production studio for a polished sound. In
the late 1980s, N.W.A popularized the West Coast rap scene with *Straight Outta Compton*.
Their lyrics reflected the poverty, racism, and violence of L.A.'s inner city and the beats
matched the rappers' rawness. See also Chapter 7 for an analysis of California imagery in
popular music.

Arcadium 2006) the geographer may leap out of his or her chair with the urge to slap an "air bass guitar," but his or her discursive analytical powers are rendered useless. Some of the band's ballads contain a loose narrative structure (e.g. "Dani California," *Stadium Arcadium* 2006), but, in general, the group has maintained a sense of ambiguity with regards to the meaning of their work. This ambiguity provides, perhaps, a challenge to the cultural geographer. On the other hand, it could be viewed as more help than hindrance. Some geographic studies of popular music have conducted discursive analyses based primarily on the lyrics of an artist, such as Moss's exploration of Bruce Springsteen's landscapes (1992). While Moss does an excellent job bringing both a materialist and feminist lens to a male rock legend's work, she does not utilize a dialectical framework to investigate both the object (e.g. "Born to Run") and the processes that lead to its creation and, in turn, the objects reproduction (Harvey 1996). Moss incorporates Springsteen's life into her analysis of his lyrics, but popular songs are more than just lyrics. Regardless of the presence or absence of a narrative structure, the sonic choices used by the artist contribute to the work. Using the aforementioned Beach Boys meet N.W.A example, instrument choice, vocal style, and the very notes and rhythm chosen produce the totality of the song. Whether or not Springsteen's guitar or Clarence Clemons's saxophone would alter Moss's findings is not the issue; she likely would have come to a similar conclusion. It should be kept in mind, however, that these choices, along with the social processes that brought Bruce Springsteen to the recording studio, as well as the processes that bring the listener to the songs, music videos, and concerts together produce the landscapes of Bruce Springsteen. All of these parts constitute the totality of the spaces and places of music (Harvey 1996).

Additionally, in a post MTV age, where a band can provide songs, music videos, documentaries, artwork, and journal entries to the fan's home computer, simply relying on the songs or albums would limit one's understanding of the overall work of art. The "webs of symbols" must be plucked not only from the songs, but from the totality of the albums, videos, and overall mythos of the band itself.[2]

L.A.'s urban form, pop culture, and postmodernity

I am in no way suggesting that L.A. has gone unnoticed in the geography of popular culture. The use of the city in Film Noir has exposed Los Angeles's darker side,

---

2   In addition to how we experience professionally produced music, new technology also seems to be changing *who* we listen to. New and (somewhat) affordable software are used to create a recording studio on one's desktop computer and internet forums like *YouTube* and *MySpace* are producing entirely new ways to consume music. Will music scenes change? Do artists still need to live in L.A., Seattle, or Nashville? What does this mean for future spaces and places of music? See Chapter 12 for insights to these questions.

thus counteracting its sunny image (Hausladen and Starrs 2005). Using noir as a stylistic framework allowed these films to express a "collective national unease" (Hausladen and Starrs 2005, 43). Additionally, McClung (1988) has discussed how literature (including noir) contrast images of utopia and Arcadia[3] against dystopia and the city. He uses Los Angeles to show the dialectical nature of such literary cities as they relate to our reading of the built landscape. Helen Hunt Jackson's 1884 novel *Ramona* managed to both challenge the treatment of Native Americans as well as shape the natural image of Southern California (DeLyser 2005). The city's Mediterranean climate and later its imported "natural" landscape (orange and palm trees as opposed to desert scrub) give it a tangible sense of an Arcadian city. Yet, literary work has typically presented the city as a "betrayed arcadia" or a "pathetic dystopia" (McClung 1988, 34). McClung uses this dialectic as a way to show how the real and the imagined are in tension in the making of a city such as Los Angeles. This idea of contextualizing art and architecture to understand how the city is imagined is relevant to this chapter. The trick, however, is to move past the notion of the utopian/dystopian binary and embrace the multiple imaginings and realities of the city.

It is my argument that the Red Hot Chili Peppers express not a national unease, nor a straightforward critique of utopia, but rather a struggle to accept the multiple realities of postmodern existence, which I would argue is a logical progression of L.A. noir. Rather than simply point out "paradise with a seamy side" (Hausladen and Starrs 2005, 60), the band embraces both worlds to show how they are not mutually exclusive. This is not a form of hypocrisy, but rather the acknowledgement of coexisting and competing realities. In turn, the band's lyrical and sonic choices are influenced by the postmodern nature of the city where they live. In other words, the Red Hot Chili Peppers are a band presenting both a "paradise" and a "seamy side" of the city. The band continues to show the underbelly of society through their openness about drug addiction, death, and despair in a noir fashion, but also addresses the global networks originating in "a Hollywood basement" ("Californication," *Californication* 1999). Rather than dwelling on the darker side of Los Angeles, the band embraces it as a necessity for the existence of the city's good side. This is not about exposing Hollywood's "lies," but about coming to terms with them and incorporating them into daily life.

Los Angeles is a city that defies modern notions of urbanism (Dear and Flusty 1998). In geographic discourse, it has been described as postmodern (Dear and Flusty 1998; Soja 1989) or even deviant (Davidson and Entrikin 2005). Los Angeles is "a peculiar composite metropolis that resembles an articulated assemblage of many different patterns of change affecting major cities in the United States and elsewhere in the world—a Houston, a Detroit, a Lower Manhattan, and a Singapore amalgamated in one urban region" (Soja et al. 1983, 195-96). This image of nontraditional urban form takes Tuan's ideas of literary London (1991) in a different direction. Not only can popular culture construct an unreal experience

---

3   A term commonly used in fiction that represents idealized, idyllic rural areas.

of the city, but also the absence of traditional urban centers and public spaces seem to produce the need for this very popular culture.

This materialist dialectic of music/place production should not, however, mask the postmodern nature of the city and the music. While Soja's socio-spatial dialectic (1980) is at work, the following analysis equally stresses the importance of Soja's work on Los Angeles as a heterogeneous city (see, for example, Soja 1996a). Soja has attempted to transcend the binary constraints of the dialectic, even proposing a "trialectic" approach while still encouraging geographers to move beyond three-part frameworks (Soja 1996b). So, while I am suggesting that the Red Hot Chili Peppers and Los Angeles are linked in a dialectic fashion, the band and the city as beings, places, and spaces transcend proposed meta-narratives in true postmodern fashion. As I will show, both are the amalgamation of competing and coexisting realities.

The band's union of "high culture" and the vernacular into their work positions them as postmodern artists (Urry 2002, 79-80). By layering an orchestra with a distorted electric guitar ("Midnight," *By the Way* 2002), the band can appeal to those with more cultural capital than economic capital (Urry 2002). By embracing juxtaposition while rejecting a meta-narrative, the band works from a postmodernist point of departure (Harvey 1990). Harvey (1990, referencing McHale 1987) describes the shift from the epistemological to the ontological in postmodernism; that is, interpreting and representing how we know the truth gives way to a world of many intersecting and diverging truths. Leaving an idealized reality out of their music, the Red Hot Chili Peppers can seamlessly attack the spread of a homogenized Western culture, while touring around the world to promote the same Western culture.

Blood Sugar Sex Magik and "Suck My Kiss"

The 1989 cover of Stevie Wonder's "Higher Ground" (*Mother's Milk*) gave the band mainstream recognition, but it was with the 1991 album *Blood Sugar Sex Magik* that the Red Hot Chili Peppers became a powerful music presence, and, I would argue, a clear postmodern turn for the band. Los Angeles was always present in earlier work (e.g. "Out in L.A." and "True Men Don't Kill Coyotes," *Red Hot Chili Peppers* 1984), but *Blood Sugar Sex Magik* more fully embraced the band's home town. The album was created in a rented Hollywood mansion where the band lived until the recording was complete. The process was captured in a documentary about the album, *Funky Monks*. This was the first time acclaimed producer Rick Rubin worked with the band (who continued to work with them on four subsequent albums), and a certain sense of maturity is evident when comparing *Blood Sugar Sex Magik* to previous efforts. While their early albums did reference the city, deliberate and tangible connections to the many realities of Los Angeles is first present on this album and the associated artwork and music videos. It is the starting point for the postmodern geographies of the Red Hot Chili

Peppers. Previous albums used pastiche to a certain extent, but at this point in their discography, postmodern play becomes tangible and appears purposeful.

The song "Suck My Kiss" (*Blood Sugar Sex Magik* 1991) is a sexually charged funk anthem, although its sonic values alone do not necessarily produce the Red Hot Chili Peppers' Los Angeles. The video, however, is a mix of footage from the *Funky Monks* film and shots from a military parade in Los Angeles. The opening shot is a black and white view of the classic "Hollywood" sign, instantly invoking geography. Next is a red duotone shot of airplanes flying in formation above festive balloons. A black and white Los Angeles of the vernacular is played against the "red and white" Los Angeles of state power. Drummer Chad Smith rides his Harley sans helmet through the Hollywood Hills, while uniformed soldiers march in a parade honoring the military; two separate images showing both disobedience and order. Next, shirtless bassist Flea and guitarist John Frusciante record this very song, bouncing and moving to the music, almost sexually; yet, almost simultaneously a tank rolls through Los Angeles behind a marching band. The American flag is waving throughout the red-hued scenes, while the only overt symbols in the vernacular shots are tribally inspired tattoos. A hint of dissent arrives at the end, when a protester in sad clown makeup holds up an out-of-shot sign. Then, several green-hued shots of Los Angeles appear.

The "Suck My Kiss" video echoes Lefebvre (1991); a (perhaps) materialist approach for a postmodern band. The state shots are "representations of space," i.e. the dominant space of order, while the vernacular shots are "representational space," i.e. "the loci of passion of action, of lived situations" (Lefebvre 1991, 42). The video, however, departs from Marxism in that it is not an obvious attack on state order and hegemony, nor is it a pure celebration of artistic emotion and humanistic freedom. Instead, it plays the two narratives off one another. They are two inherently different social scenes, and yet both rooted in the same city. The streets of L.A. exist for a long-haired drummer to ride his motorcycle as well as a venue to pledge allegiance to the state. The place is a product of both (or more) spaces, and these spaces could not exist without such a place. In many ways the video expresses the conflicting use of the postwar American road. While designed for military purposes and constructed from the engineer's pragmatic vision (Brown 2005), it has been used in subversive ways as a means for subcultural (often male) escape (Cresswell 1993).

"Under the Bridge"

Arguably the band's most recognized song, "Under the Bridge" (*Blood Sugar Sex Magik* 1991) is not only set in Los Angeles, but uses the city as a driving character. The song deals with lead singer Anthony Kiedis's addiction to heroin, and was "discovered" in his stack of poems by producer Rick Rubin (see the *Funky Monks* documentary). It is a dark song, opening with the emptiness and solitude Kiedis felt at the depth of his addiction:

Sometimes I feel
Like I don't have a partner
Sometimes I feel
Like my only friend
Is the city I live in
The city of Angels
Lonely as I am
Together we cry
("Under the Bridge" *Blood Sugar Sex Magik* 1991)

The city is the only "person" in his life that understands who he truly is, who can see through his demons to his true soul. "I walk through her hills/'cause she knows who I am/She sees my good deeds/ and she kisses me windy ("Under the Bridge," *Blood Sugar Sex Magik* 1991)."

The video furthers the concept of the personified city. The downtown skyline is played against the floating bodies of the band members, almost making it the fifth 'Chili Pepper. As Kiedis walks through downtown Los Angeles, weaving through citizens and streets, he sings the lyrics to the camera, as if he is introducing the viewer to his friend L.A. He interacts with local merchants to show a connection with the bodies that make up the city. The fact that this is downtown, and not Santa Monica or Pasadena, embraces the vernacular side of the band's tastes. Graffiti, non-Anglo Angelinos, and modest architecture help to highlight both the despair that runs throughout the city, as well as the urban "survivors" that occupy the city. While pastiche is not a driving force of the song or video, it is a testament to the role Los Angeles has played for the Red Hot Chili Peppers.

"By the Way"

The video for "By the Way" (*By The Way* 2002) is more obviously Los Angeles than the song itself. It opens with a shot of an Echo Park Avenue street sign,[4] then tilts down to Anthony Kiedis hailing a cab. This video is the most interesting of the band's video catalog as it does not revolve around a musical performance, a collection of images, or a fantastic premise. While the video content is centered on a fictional event, the viewer follows a narrative of the band as one might actually see them at home in Los Angeles. The video has a voyeuristic style, which gives it a more "authentic" feel than the interactions seen in "Under the Bridge."

An obsessed fan drives a cab that Kiedis hails. The driver slides a Red Hot Chili Peppers CD (the "By the Way" single) into the stereo, which Kiedis politely acknowledges. As the song's intensity progresses, so too does the driver's excitement. The cab begins racing through L.A., the cabbie pounding on his

---

4   The Echo Park and Silverlake neighborhoods of Los Angeles are known for their bohemian vibe as well as recent gentrification efforts.

steering wheel along with the funk, and Kiedis grows uncomfortable. He pulls out his phone to call for help, but the crazed driver slams on the brakes to knock it out of his hands. The driver pulls into a deserted tunnel and stops the cab. While he pulls out flares from the trunk, Kiedis takes out a PDA (personal digital assistant) (two phones—how L.A. is that?).

Sitting at a sidewalk café somewhere in the city, Flea and John Frusciante receive a text message from the now abducted Kiedis. Flea's hair is electric blue: a punk rock juxtaposition with the bourgeois setting. They laugh at the apparent kidnapping joke, and go back to their drinks. Meanwhile, Kiedis is subjected to the cabbie dancing in front of the cab with lit flares in hand, performing what can best be described as a "white-trash island dance." He sends another text and this one is taken seriously.

The rescue vehicle is absurd; a yellowish-orange 1960s vintage Ford Bronco, which is a perfect complement to the equally absurd rescue mission. Flea and Frusciante chase the cab, which is once again driving recklessly through Los Angeles. Kiedis breaks the rear window of his prison, climbs onto the roof of the cab and leaps into the Bronco. The two vehicles part ways, and the final shot is of the cab now picking up the unsuspecting 'Chili Peppers drummer Chad Smith.

Beneath the sudden narrative structure in the band's work, there are significant underlying images in the video. Once again, the streets of Los Angeles are used, but unlike the L.A. of "Under the Bridge," this setting is not bound to the vernacular. The story begins in the gentrified Echo Park with Anthony Kiedis dressed in a hip blazer. This is quite the departure from strolling though downtown in a ratty T-shirt and long hair. While it might look like the band has forgotten its roots with this journey into elite Los Angeles, the Red Hot Chili Peppers have actually achieved a new level of pastiche. Refined Echo Park is matched with a wild car chase. Respectable citizens meet deviant cab drivers. Trendy rockstars drink tea at a sidewalk café then drive a monster truck. The video moves from clean streets to forgotten tunnels revealing an underworld of trash and graffiti. And while this dialectical Los Angeles is played with, the car chase event evokes Hollywood action hero machismo. All the while, the music shifts from the haunting high range of Keidis's voice and steady, quiet drumming to the heavy bass and nonsensical lyrics, then back again. The song seamlessly shifts from "pretty" to "rockin'." Whatever the influence, the content of this video is the sum of the city. The video further produces a Red Hot Chili Peppers vision of Los Angeles. Flea's blue hair set against a luxurious café might look odd against the traditional milieus of Chicago or New York, but it actually makes sense in L.A. Images like those in the video construct the Los Angeles that allows and produces such juxtaposition.

"Tell Me Baby"

The Red Hot Chili Peppers' latest work, *Stadium Arcadium,* is an attempt not to be confined to the boundaries of Earth, which is evident from the celestial theme of the double album. The band may have tried to produce a universal album, but it is still rooted in Los Angeles. "Tell Me Baby" (*Stadium Arcadium* 2006) is a song about migrating to Hollywood in the hope of making it big. It tells the story of every wannabe rockstar, actress, and model that has ever moved to the city:

> They come from every state to find
> Some dreams were meant to be declined
> Tell the man what did you have in mind
> What have you come to do
> ("Tell Me Baby," *Stadium Arcadium* 2006)

The very subject of the song is a mixture of elites and commoners. After all, these worldly rockstars were once those trying to make it big themselves. The video furthers that concept, shot entirely in a small dingy room with a simple camcorder. Various musicians and singers are brought in front of the camera as if they are auditioning during the song's intro. Flea appears on the tiny stage and launches into the music when the song picks up. Shots of the Red Hot Chili Peppers playing the song are mixed with the wannabes covering the same one. Some have potential, others are "meant to be declined." As the video progresses, the band sneaks onto the stage with the auditioning musicians and begins playing with them. The elite rockstars and the lowly wannabes switch physical and social roles; nobodies sing with celebrities backing them up. The video culminates with everyone playing, dancing, and singing in a frenzied jam session. The energy and excitement of the wannabes merges with the sheer joy of the band to great effect.

Through the lyrics, the band acknowledges the palimpsest that is Los Angeles:

> This town is made of many things
> Just look at what the current brings
> So high it's only promising
> This place was made on you
> ("Tell Me Baby," *Stadium Arcadium* 2006)

By singing that L.A. is "made of many things," the band brings together the different worlds of Los Angeles that they have lived and encountered throughout their career. It acknowledges the capitalist structures that draw artists from across the world, as well as the rebellious nature of the artists themselves. Through "Tell Me Baby" the listener/viewer learns that L.A. is hope, L.A. is despair, L.A. is good, L.A. is bad.

The importance of this particular video is stressed in the 4 June 2006 "Fleamail," occasional open e-mails written by Flea to fans in a punctuation-free stream of consciousness, evocative of Jack Kerouac:

we made a video for the song 'tell me baby' and we just saw it yesterday i am
so excited about it it is the best one we ever did, it was done by dayton and faris
they made us the best video of our career and i cant say what it is and ruin the
debut of it, not to sound so high and mighty like it is such an important thing
but, i think it is the most beautiful piece of film that has ever represented us
(Balzary 2006).

Conclusion

In Mike Davis's less-than-flattering critique of the city (1990), he touches upon the
fact that "Los Angeles is usually seen as peculiarly infertile cultural soil, unable
to produce, to this day, any homegrown intelligentsia" (17). For a Marxist scholar,
such an elite stance is surprising, as it seems to betray his apparent motives. He
attacks Los Angeles for not mimicking "old world" urban life *and* for not breaking
down the norms of class structure. At the end of his history of intellectual Los
Angeles, Davis mentions gangsta rap as a conduit for the marginalized voice of
the inner city, although he criticizes the West Coast's rappers for "selling out."
They are not making music as an art form, but are instead focused on making
money. No doubt this concerns a Marxist like Davis, yet he fails to remove his own
bias towards the capital structures that rule Los Angeles. How can anyone truly
criticize a person born into the poverty and violence of Compton who wants to
make money to escape it? By accepting the juxtaposition inherent in being true to
one's art while making money, of fighting existing structures while utilizing them
to create music, the Red Hot Chili Peppers have been able to prove that rejecting a
meta-narrative in favor of accepting multiple narratives is possible.

Forman deftly discusses the territoriality of rap and its "strong local allegiances"
(2000, 88) while avoiding ivory tower judgment. Rappers explicitly talk of the
local through their lyrics, giving voice to the male aggression found in the 'hoods.
The Red Hot Chili Peppers, not born and raised in the ghetto, but by no means elite
Angelinos, give voice to their locale. For the band, Los Angeles is a city of tension
and hypocrisy. The band members are of a world that acknowledges the anger of
Compton, while not discounting the opulence of the West side.

However, the Red Hot Chili Peppers depart from the localism of rap in that
they weave multiple local identities into their artwork, video images, and music. It
is rare to hear a straightforward critique of the city in the lyrics. It is through such
subtlety that the band is able to create pastiche, interpretive artistry, and to address
the totality of their city. Rather than limiting themselves to one neighborhood of
the sprawling city, or to themes of male aggression and poverty, or fashion and
leisure, that is, by rejecting ontology, they are actually able to produce a version of
L.A. that speaks to all classes and identities. Conversely, if we accept the idea of
Los Angeles as a postmodern city (Dear and Flusty 1998; Soja 1996b) not fitting
with classic models of urban life, it makes sense that L.A. would produce such a
postmodern band. The Red Hot Chili Peppers have made a career out of pastiche;

borrowing and blending elements of rap, rock n' roll, funk, punk, country, and jazz. Songs from Hank Williams are skillfully reinterpreted with bass lines reminiscent of Bootsy Collins ("Why don't you love me?" *Red Hot Chili Peppers* 1984). And while it is doubtful that this music could be made in any other city (keeping in mind that *One Hot Minute*, the band's most criticized post-*Blood Sugar Sex Magik* album, was recorded in Hawai'i), the Red Hot Chili Peppers' sonic choices and images, like that of a blue-haired Flea sitting at a refined sidewalk café, have helped to produce an untraditional image of L.A.

This chapter is by no means the endpoint of studying the music of the Red Hot Chili Peppers. While preceding geographic work has focused on place, the spaces of love, lust, despair, addiction, machismo, and maturity are ripe for geographic analysis. Aitken and Craine (2002) have already begun to lay the foundation of such work. An analysis of both the places and spaces of the Red Hot Chili Peppers will be necessary to grasp the totality of the dialectical and postmodern relationships between the band and L.A. geography. And while Los Angeles has played an important role in the band's work, who knows what places and spaces the band has yet to explore?

> If you let go and let this music take you by the hand it will take you flying through skies of sound. It will zoom you up well above outer space and it will show you around planes of existence that do not share the laws and conditions of this reality. And when it brings you down to earth it will dig deep into that shit. It will also teach you to fall back without landing on your ass and to fall forward without falling on your face. Let go and you can be two places at once, you can be as big as the whole universe and as small as an atom simultaneously … In the words of one of the supreme gods of funk, "Nothing is good unless you play with it." The Red Hot Chili Peppers … have played with light, darkness, sound, silence, form, air, and space to make music that plays with the listener (Frusciante 2006).

References

Aitken, S.C. and Craine, J. (2002), "The Pornography of Despair: Lust, Desire, and the Music of Matt Johnson," *Acme: An International E-Journal for Critical Geographies* 1:1, 91-116.

Balzary, M. "Flea." (2006), "Fleamail – June 4 2006." <http://www.redhotchilipeppers.com/news/journal.php?uid=241>, accessed 16 April 2007.

Bell, T.L. (1998), "Why Seattle? An Examination of an Alternative Rock Culture Hearth," *Journal of Cultural Geography* 18:1, 35-47.

Brown, J. (2005), "A Tale of Two Visions: Harland Bartholomew, Robert Moses, and the Development of the American Freeway," *Journal of Planning History* 4:1, 3-32.

Carney, G.O. (ed.) (1978), *The Sounds of People and Places: A Geography of American Music from Country to Classical and Blues to Bop* (Lanham, MD: Rowman & Littlefield).

—— (1995), *Fast Food, Stock Cars, & Rock-n-Roll: Place and Space in American Pop Culture* (Lanham, MD: Rowman & Littlefield).

Connell, J. and Gibson, C. (2004), "World Music: Deterritorializing Place and Identity," *Progress in Human Geography* 28:3, 342-61.

Cresswell, T. (1993), "Mobility as Resistance: A Geographical Reading of Kerouac's 'On the Road'," *Transactions of the Institute of British Geographers* 18:2, 249-62.

Davidson, R.A. and Entrikin, J.N. (2005), "The Los Angeles Coast as a Public Place," *The Geographical Review* 95:4, 578-93.

Davis, M. (1990), *City of Quartz: Excavating the Future in Los Angeles* (New York: Verso).

Dear, M. and Flusty, S. (1998), "Postmodern Urbanism," *Annals of the Association of American Geographers* 88:1, 50-72.

DeLyser, D. (2005), *Ramona Memories: Tourism and the Shaping of Southern California* (Minneapolis: University of Minnesota Press).

Ford, L.R. (1971), "Geographic Factors in the Origin, Evolution, and Diffusion of Rock and Roll Music," *The Journal of Geography* 70, 455-64.

Ford, L.R. and Henderson, F.M. (1978), "The Image of Place in American Popular Music: 1890-1970," in G.O. Carney (ed.).

Forman, M. (2000), "'Represent': Race, Space, and Place in Rap Music," *Popular Music* 19:1, 65-90.

Frusciante, J. (2006) "Biography – Stadium Arcadium." <http://www.redhotchilipeppers.com/news/bio.php>, accessed 8 April 2007.

Gibson, C. and Dunbar-Hall, P. (2000), "Nitmiluk: Place and Empowerment in Australian Aboriginal Popular Music," *Ethnomusicology* 44:1, 39-64.

Gill, W.G. (1995), "Region, Agency, and Popular Music: The Northwest Sound, 1958-1966," in G.O. Carney (ed.).

Graves, S. (1999), *A Historical Geography of the Music Industry* Unpublished doctoral thesis. Department of Geography, University of Illinois at Urbana-Champaign.

Greene, P. (2001), "Mixed Messages: Unsettled Cosmopolitanisms in Nepali Pop," *Popular Music* 20:2, 169-87.

Gumprecht, B. (1998), Lubbock on everything: The evocation of place in popular music (a West Texas example). *Journal of Cultural Geography* 18(1), 61-82.

Harvey, D. (1990), *The Condition of Postmodernity* (Malden, MA: Blackwell).

—— (1996), *Justice, Nature, and the Geography of Difference* (Malden, MA: Blackwell).

Hausladen, G.L. and Starrs, P.F. (2005), "L.A. Noir," *Journal of Cultural Geography* 23:1, 43-69.

Hudson, R. (2006), "Regions and Place: Music, Identity, and Place," *Progress in Human Geography* 30:5, 626-34.

Kong, L. (1995), "Popular Music in Geographical Analyses," *Progress in Human Geography* 19, 183-98.

Lefebvre, H. (1991) [1974], *The Production of Space* (Oxford: Blackwell).

Leyshon, A., Matless, D., and Revill, G. (1995), "The place of music," *Transactions of the Institute of British Geographers* 20:4, 423-33.

Mitchell, D. (2000), *Cultural Geography: A Critical Introduction* (Malden, MA: Blackwell).

McClung, W.A. (1988), "Dialectics of Literary Cities," *Journal of Architectural Education* 41:3, 33-7.

McHale, B. (1987), *Postmodernist Fiction* (London: Routledge).

Moss, P. (1992), "Where is the 'promised land'? Class and Gender in Bruce Springsteen's Rock Lyrics," *Geografiska Annaler Series B, Human Geography* 74:3, 167-87.

Scott, A.J. and Soja E.W. (eds.) (1996), *The City: Los Angeles and Urban Theory at the End of the Twentieth Century* (Los Angeles: University of California Press).

Soja, E.W. (1980), "The Socio-spatial Dialectic," *Annals of the Association of American Geographers* 70:2, 207-25.

—— (1989), *Postmodern Geographies: The Reassertion of Space in Critical Social Theory* (New York: W.W. Norton).

—— (1996a), "Los Angeles, 1965-1992: From Crisis-generated Restructuring to Restructuring-generated Crisis," in A.J. Scott and E.W. Soja (eds.).

—— (1996b), *Thirdspace: Journeys to Los Angeles and Other Real-and-imagined Places.* (Cambridge, MA: Blackwell).

Soja, E.W., Morales, R., and Wolff, G. (1983), "Urban Restructuring: An Analysis of Social and Spatial Change in Los Angeles," *Economic Geography* 59:2, 195-230.

Stump, R.W. (1998), "Place and Innovation in Popular Music: The Bebop Revolution in Jazz," *Journal of Cultural Geography* 18:1, 11-34.

Tuan, Y. (1974), *Topophilia: A Study of Environmental Perception, Attitudes, and Values* (Englewood Cliffs, NJ: Prentice-Hall).

—— (1991), "Language and the Making of Place: A Narrative-descriptive Approach," *Annals of the Association of American Geographers* 81:4, 684-96.

Urry, J. (2002), *The Tourist Gaze: Leisure and Travel in Contemporary Society*, 2nd Edition. (London: Sage).

Valentine, G. (1995), "Creating Transgressive Space: The Music of kd lang," *Transactions of the Institute of British Geographers* 20:4, 474-85.

Zelinsky, W. (1999), Review of *The Place of Music* by A. Leyshon, D. Matless and G. Revill *Economic Geography* 75:4, 420-22.

Discography

Red Hot Chili Peppers (1984), *Red Hot Chili Peppers* Compact Disc. EMI.
Red Hot Chili Peppers (1989), *Mother's Milk* Compact Disc. EMI.
Red Hot Chili Peppers (1991), *Blood Sugar Sex Magik* Compact Disc. Warner Brothers.
Red Hot Chili Peppers (1995), *One Hot Minute*. Compact Disc. Warner Brothers.
Red Hot Chili Peppers (1999), *Californication* Compact Disc. Warner Brothers.
Red Hot Chili Peppers (2002), *By the Way* Compact Disc. Warner Brothers.
Red Hot Chili Peppers (2006), *Stadium Arcadium* Compact Disc. Warner Brothers.

Chapter 10

The Geography of "Canadian Shield Rock": Locality, Nationality and Place Imagery in the Music of the Rheostatics

Olaf Kuhlke

Dave Bidini, Martin Tielli, Tim Vesely and Don Kerr are a big part of a musical revolution that proves that rock music made in Canada doesn't have to sound like it came from somewhere else (Brown 1996).

Introduction: Nation, society and the place of the Rheostatics

On 30 March 2007, an era in the history of Canadian alternative rock music came to an end: The Rheostatics, one of the country's most beloved bands, performed their last concert at Toronto's Massey Hall, after more than 20 years of recording and touring together. While individual members of the group continue to make music, and several of them have successful solo careers already in place, fans and critics are mourning this loss extensively (Hunt 2007; Rayner 2007). As one of Canada's most influential alternative rock bands, the Rheostatics' music, their concert appearances, and their witty social commentary have always been reflective of growing up and being a young adult in Canada (Doucet 1998). But even more so, they have extensively engaged with the often problematic and contested notions of regional and national identity (Millard et al. 2002). Fiercely patriotic, yet often parodying Canadian culture and traditions, their music has created distinctive representations of what Canada is to them and to the wider public, and what it ought to be. In this chapter I demonstrate how the group and its individual members, through music, writing and artwork, have reaffirmed four distinct yet complementing notions of Canadian national identity, and how they have particularly utilized nationalist place imagery to do so.

First, I examine how they reaffirm Canadian national identity as distinctly multi-cultural and multi-lingual, even beyond the notions of how traditional English-French relations have shaped the Canadian experience. In the past, geographers have paid particular attention to how Canada's many indigenous and non-indigenous groups have searched for and struggled over the creation of a single nation-state (Ayres 1995; Kaplan 1994a, 1994b; Wilson 2005). Many Canadian artists have creatively reflected on the often uneasy relationship between indigenous people, French Canadians, and immigrants of other European origins (Beaudreau 2002; Hill 1995; Lang 2004; Wright 2004). Second, I show how the

Rheostatics creatively engage with the popular notion of defining Canada through what it is not—namely the United States. This negative definition of nationalism has repeatedly captured the Canadian imagination, and continues to be a powerful medium of establishing cultural boundaries between these two North American countries today (Reich 1999). Such "We-Are-Not-American"-ism is based on the often deep-seated resentment of losing a sense of Canadian identity due to the cultural influence of the United States. George Grant already expressed such lament in the mid-1960s, and his work continues to be influential today:

> To be a Canadian was to build, along with the French, a more ordered and stable society than the liberal experiment in the United States. Now that this hope has been extinguished, we are too old to be retrained by a new master ... The element necessary to our existence has passed away (Grant 2005, 5).

Such overt definition of Canadians as "not American" pervade popular culture and music, and was perhaps best captured in a beer commercial for the Molson brewery that aired in 2000 on Canadian television, right in time for the Stanley Cup playoffs. In it, a Canadian named "Joe" steps is front of a giant video screen and proclaims:

> Hey. I'm not a lumberjack, or a fur trader ... and I don't live in an igloo or eat blubber, or own a dogsled ... and I don't know Jimmy, Sally or Suzy from Canada, although I'm certain they're really, really nice. I have a prime minister, not a president. I speak English and French, not American, and I pronounce it about, not a boot. I can proudly sew my country's flag on my backpack. I believe in peace keeping, not policing. Diversity, not assimilation, and that the beaver is a truly proud and noble animal. A toque is a hat, a chesterfield is a couch, and it is pronounced zed, not zee. Zed. Canada is the second largest landmass, the first nation of hockey, and the best part of North America. My name is Joe, and I am Canadian (Molson 2000).

In this chapter, I will show that the Rheostatics have created similar representations of Canada as an opposite to the United States, by creating a variety of imaginary musical landscapes that parody the United States and define Canadian society as diametrically opposed to America.

Third, the Rheostatics have continuously reaffirmed in the creation of their "soundscapes" the notion of a successful Canadian experiment with democratic socialism and many of its components. Despite the current shift towards conservative leadership in the federal government, Canadian politics has long been dominated by democratic socialist ideas such as universal health care, strong support of the educational system, stricter gun control, and a more extensive welfare system than in the United States (Buckstein 2005; Jackson 2005; Whitehorn 1992). In fact, such trust in government, as opposed to the fundamental American credo of limited government is also expressed in the third theme I will explore here. One of the founding principles of the Canadian confederation in 1867 was "peace, order and good government" (Dyck 2000). Beyond the legalistic interpretation of this tripartite term that regulates federal and provincial powers in Canada, it is often used as a broad characterization of a political culture that emphasizes the high degree of deference to the law that Canadians have, and their trust and expectations in the federal government (Lipset 1991). In their music, the Rheostatics pay tribute to the many ways in which Canadians trust their government in providing them with social assistance, and they openly resist the recent changes to this system brought about by more conservative federal and provincial governments.

Last, and perhaps most importantly, Canadian national identity is often evoked through powerful metaphors of nature. Our modern nation-states were created over extensive periods of time, and many individuals contributed with their ideas to the economic, social and political structures that characterize modern nations. In order to create a sense of collective belonging and self-representation, nations have traditionally used a series of symbols to express what binds them together (Anderson 1991; Bhabha 1993; Halbwachs 1985; Hall 1994). In times of change, crisis, or instability, nations have utilized natural symbols, including landscape art, to tell the story of a nation, and to represent their wish for continuity and durability (Linke 1999). This particularly applies to Canada (Desbiens 2004; Hill 1995). The constant threat of the possible secession of Quebec and the continued struggle to maintain national unity have created a survival mentality in Canada that the linguist Northrop Frye referred to as the "garrison mind" (Lecker 1993). Such notion is thus often reaffirmed by representing Canadians as hardy, tough, and in tune with nature. Metaphors of nature naturalize identity, move it away from contestation, and bind it symbolically to the land (Linke 1999; Kuhlke 2004). The Rheostatics, as I will show, repeatedly utilize landscape imagery and reference to the literal and figurative "nature of Canada" to express their sense of patriotism and belonging, and create an audible and visual landscape of Canada that celebrates nature.

Methodologically, my analysis draws on a discursive reading (Gee 2005; Lees 2004) of the lyrics and music of selected Rheostatics albums and songs, and the events and performances they were involved in. In particular, I place the artistic work of the band in the context of the four discourses, or debates about Canadian nationalism, to show how the band operates within them and at times contradicts or actively challenges these affirmations of collective identity.

Chanson les ruelles: Celebrating Canada's multilingual mosaic

Canada's effort to construct a truly multicultural and multilingual society has been marked by a checkered past and seemingly never-ending quarrel between separatist groups and the rest of the nation. Prior to 1967, when the Canadian immigration law was fundamentally overhauled, the government had in many ways enforced the preferred immigration of European settlers with its "White Canada" policy, and actively worked against the large-scale migration of people of Asian and African descent (Ward 2003). While Americans were incarcerating Japanese immigrants in labor camps during World War II, the Canadian government conducted a much less publicized campaign against its own populations. Thousands of Japanese, along with conscientious objectors, were sent to work camps in the Canadian Rockies to assist with the construction of roads that now connect the national parks visited by millions each year in Alberta and British Columbia (Waiser 1995). Ever since these days of discrimination formally ended in 1967, the Canadian ethnic mosaic has been fundamentally altered. In the late 1960s, almost 70 percent of the immigrant force coming to Canada annually was of European origin. By 2001, more than 60 percent of the arriving new immigrants came from Asia (Statistics Canada 2001). Canada has become a multiethnic society in which minority groups can strongly maintain their sense of collective identity and foster their language. While such transition to a more diverse society has not been without resistance and resentment, perhaps the larger issue in the creation of a unified Canadian national identity has been the continued threat of secession of Quebec, and the social and cultural issues associated with it (Young 1998).

Throughout the late twentieth century, several referenda were held in Quebec to achieve the independence of the province from the rest of Canada; but none of these succeeded. The last one of these votes was held in 1995, and independence defeated by only a very small margin (Clark and Kornberg 1996). Quebecois secessionists have always argued that Quebec is a distinct society with a separate culture and language, but the federal government has never fully acknowledged this in the Canadian constitution. By and large, the debate about the role of Quebec within Canada is one about the relationship between the federal government and provincial powers, and the right to provincial self-determination of certain policies such as language education (Bernard 1978; Conway 2004). Canadian federalists have repeatedly attempted to maintain the unity of Canada by making it an officially bilingual country, by negotiating many concessions to the French-speaking province in an effort to quell debate about the separation of the country into two, and by warning of the dire economic consequences of a divided country (Young 1994). In addition, the Canadian government spent considerable resources after the failed 1995 referendum to advocate Canadian unity by running an image campaign portraying the benefits of a Quebec within Canada (Stevenson 2004).

Subsequently, popular imaginations of Canadian national identity, as produced by English-speaking Canadians, often reaffirm the multi-ethnic, and bilingual character of the nation, and have repeatedly perpetuated the myth of harmonious unity (Duffy 2005). Just as the Molson Canadian beer commercial of 2000 sought to de-emphasize tension within the nation and set it apart against the United States by highlighting biliguality and a peaceful mosaic of cultures, the Rheostatics celebrate this myth in their music. As an Anglo-Canadian band from Etobicoke, a Toronto suburb, the Rheostatics particularly evoke notions of multilingual harmony in the songs *"Chanson Les Ruelles"* and *"Motorino."* The song *"Chanson Les Ruelles"* ("Song of the [little] Streets") was written in French, and it translates as follows:

I've decided to write this song in French. It's not the words, it's the melody that speaks with clarity. In a country where there are two languages, there are lots of people with guitars, in the church basements and backyards, on the radio and in the bars. Perhaps our style will suffer in the United States. But we sing for ourselves as everyone knows. Let the young rock stars go south. They'll forget and we will too. It's not their fault, it's a fact of the world. I hope the border will not disappear (Rheostatics 1991).

Here, the Rheostatics clearly state that while not all Canadians may speak French, or comprehend all the words of the song, it is nevertheless the melody, the intonation, the musical quality of the French language that all Canadians understand and accept as part of their identity. They situate Canada as a bilingual country, where French may not be understood clearly by all, but may be supported and conveyed in its beauty by the power of instrumentation, more specifically the guitar. The affirmation of bilingualism is also used to separate Canadian national identity from the United States. The song implies that singing in French will make the group less popular in the United States, and their "style will suffer" (Rheostatics 1991). The statement "we sing for ourselves" is one that can be metaphorically read as not just referring to the band itself, but to the whole of Canada. By singing in both French and English, Canadians truly seek to sing for themselves, not for the much more commercial market south of their border. Bands who migrate to the United States, according to this song, will be soon forgotten as being Canadian bands, and they will surrender their identity by crossing over to the United States. While the Rheostatics lament this as a common incidence, they nevertheless express the hope that the border will remain and the distinct cultural character of Canadian music will linger. In their cultural inclusiveness, their music is portrayed as distinctly Canadian. They thus engage in what Smith (1991) calls the creation of national identity as a territorialized myth. Canada, by virtue of its distinctness as an officially bilingual cultural unit, is celebrated as a territory with recognizable boundaries. In the celebration of this cultural diversity they are perpetuating the myth that Canada is truly bilingual, despite the fact that in 2001 only 18 percent of the total population actually spoke both English and French fluently. This, in fact, is similar to the percentage of the US population that is bilingual, and speaks at least one more language than English (CensusScope 2007).

With this and other songs, the Rheostatics, like many other Canadian bands, are thus fully engaged in the reenactment and representation of national identity as territorialized myth. They project wishful thinking about characteristics of national unity into an imaginary musical landscape in which the problematic dichotomy between French- and English-speaking Canadians does not exist. Another song perhaps even more forcefully illustrates how the band creates linguistic pastiches to playfully engage with the cultural hybridity of Canada. In the song "*Motorino*," the band creates imaginary landscapes and scenarios that highlight the many ethnic heritages that have contributed to the Canadian mosaic:

Donna mio!! Sacrosancti! Il nostro mondo interno e simplice. Monto Baldo e un sasolin. O! Face Bella! The pain of feeling too much and knowing too little. The autobahn on a motorino. The fear of being attacked, the punching too soon. The Canoe that cruised the moon. The unmovable moon. Far above the ravens. In silent shifting blue. In my flying canoe. It's a sign (Rheostatics 1996).

Here, lead singer Martin Tielli simultaneous recalls his Italian heritage and the artistic influence of indigenous Canadian stories on his songwriting. He creates a patchwork of dream sequences centered around a motor scooter (*Motorino* is a popular brand of Canadian electric motor scooters modeled after the famous Italian *Vespa*). What's important here is not that we clearly understand this dream and its exact meaning of it. But rather, we can understand this verse as a spontaneous, imaginative celebration of cultural hybridity. This song symbolizes how Tielli sees himself as a Canadian influenced by multiple cultural traditions; first, his Italian heritage, as expressed by the fact that part of the lyrics are written in Italian; and second, the song makes reference to the canoe, raven and moon, all prominent symbols of indigenous Canadian mythology and spirituality. While none of the band members of the Rheostatics has native Canadian ancestors, Martin Tielli acknowledges here how Canadians have celebrated representation of indigenous cultures as an integral part of Canadian identity. In fact, indigenous symbols have been significantly co-opted by both Anglo- and Franco-Canadian political elites to suggest a harmonious coexistence of native and non-native ethnicities in Canada (Sachdev et al. 2006). Despite the fact that native Canadians have endured (and continue to experience) segregation and racism, and still lag behind significantly in economic terms (Yalnizyan 2000), the evocation and celebration of an idealized egalitarian cultural hybridity is fairly common in the Canadian music scene (O'Connor 2002; Obert 2006), and the Rheostatics are an integral part of this process. With their idealization of a Canadian cultural hybridity that de facto does not exist, the Rheostatics thus aestheticize popular elite representations of nationalism that postcolonial theorists have so much lamented (Bhabha 1993; Ho 2000; Jackson 2005; Linke 1999).

Ranting loudly: Canada is … not America!

Postcolonial theories of nationalism have also often emphasized that expressions of nationalism primarily work through the process of "othering;" the representation of a nation's positive identity by the stigmatization of an "other," a negative oppositional force (Ho 2000; Said 1978). In the Canadian case, there is no more dominant "other" to Canada in the popular imagination than the United States of America. Canadians are preoccupied with concerns about their own sense of collective identity. Since a collective identity that includes Quebec has been hard to forge, and since the cultural and economic influences of the United States have been so overwhelming, Canadians have often resorted to the process of "othering" their identity by defining what they are not. Unlike their quintessential opposite, the United States, Canadians like to highlight that they do not have the same high rates of gun-ownership and gun-related violent crime, that Canada has a more extensive and egalitarian health care system, that Canadians are more politically astute, have a significantly higher participation in the electoral process, and provide immigrants with more opportunities to become part of mosaic of cultures rather than expecting them to assimilate to a presumably all-American value system (Thomas 2000). As such, discernable anti-Americanism is observable in the Canadian popular media, fuelling the representation of Canada as better than the United States because of the things that it does not have. One needs only to look at the recent onslaught of beer commercials sponsored by the Molson Brewery. From 2000 to 2005, the brewery has produced numerous commercials, such as the one previously cited, utilizing the "I am Canadian" slogan to market their most popular beer, Molson Canadian. In many of these commercials, Canadians are portrayed as enduring some kind of incident of American ignorance, thus rendering them "not ignorant" and smarter than their counterpart (Anderson 2007). In response, the ridiculed Canadians retaliate with unexpected force, and often violently. In the final analysis, these commercials portray Canadians as better than Americans by giving them the upper hand in the end—but only after it has been pointed out that Canadians are "not like Americans." Such "othering," as it appeared frequently in commercials, actively seeks to undermine the "myth of diffidence," the idea that Canadians are passive, lack self-confidence, and feel subordinate to the United States (Millard et al. 2002). As a consequence, many Canadian bands have engaged in boisterous celebrations of their "Canadianness," and highlighted their imagined superior status by emphasizing the biggest faults and flaws of American society—most importantly the American obsession with guns and violence, and the overt commercialism of American popular culture (an idea also explored by American filmmaker Michael Moore in *Bowling for Columbine*).

The Rheostatics fully engage with this critique of America in several of their songs. Their anti-American sentiment surfaces in the already discussed "*Chanson Les Ruelles*," but even more so in "Guns and RDA (Rock! Death! America!)." In "Guns," the Rheostatics make the connection between getting used to guns in childhood, and the consequences that such familiarity may have later in life:

> You take a gun and shoot an animal. Say like your neighbor or a squirrel. Safe to say that at any one time both can be spotted in the locale and struck down with ease. You can rest easy in knowing that the pencil rains will lodge in the target you shot at, and maybe even in some stuff you didn't shoot at. Neat. Stupid. Disgusting. You can take this recipe to the bank and put it in a long-term deposit, because the results are always guaranteed: 1. Some goof produces said weapon from a factory. 2. Some goof owns a store that peddles death. 3. Some goof buys the power toy. 4. Said last goof gets a novel idea. 5. Someone, not necessarily a goof, dies. Guns (Rheostatics 1992).

Here, the production, sale and availability of guns in general in criticized, and the process of acquiring a gun is described. The song highlights the ease with which a gun changes users—from the producer to the seller to the final consumer. In comparison, Canadian gun ownership is severely restricted. The purchase of hunting rifles requires the completion of a firearms safety course and the ownership of a firearms license. Handgun ownership in Canada is only allowed if individuals can prove that such weapon is necessary to protect their safety. Thus, private handgun ownership is extremely rare, compared to the United States. The song "Guns" thus clearly bespeaks a situation that is not common in Canada, but occurs frequently in the United States. One need only recall the recent shooting at Virginia Tech, or the Columbine High School massacre in 1999, when the debate following the shootings revealed how easy the perpetrators acquired their weapons. The song implicitly emphasizes an American issue and problem, and criticizes Canada's southern neighbor for its fascination with guns. In contrast to offering this subtle critique of what is considered a mostly American social issue, the Rheostatics are much more outspoken about another fundamental Canadian anxiety: The fear of being mistaken for an American (Jacobson 2004). Despite the many popular representations of Canadians as not acting, thinking and consuming like their southern neighbor, they are often mistaken as Americans, due to the many influences of American popular culture (Salloum 2004). The Rheostatics engage with this issue in their popular song "RDA (Rock! Death! America!)," which they played as the very last song in their last live show on 30 March 2007. As such, it holds special significance to the band and its fans, especially because it reaffirms their sense as a quintessentially Canadian band. In the lyrics, they clearly refer to the issue of being mistaken for Americans:

> Someone said we sounded like the Replacements, but we'd never be the Beatles or Byrds. Someone said we shoulda stayed in the basement instead of littering our noise on the Earth. But the dinosaurs are dying each day. They're gonna wish they never got up to pray. 'Cause I don't need this, and I don't need that. And I don't need this, and I don't need that. All my cousins live here fat off the land. I hear them lowing but I can't understand. And in the line-up where their souls can be sold, they've never heard of this Canadian band. But the dinosaurs are dying each day. They're gonna wish I never go up to play. "Hey Dave!" What?! So you say you think that you've got the answer?" I can't wait, gimme a break. I'm rabbit p-p-punching but the skin won't break. I'm gonna get a van and drive it to Graceland. I'm gonna firebomb the crowd at the gates. I'm gonna watch it from the back of the mansion. I'm gonna dig him up and lie in his ... grave. Rock! Death! America! (Rheostatics 1992).

Here, the Rheostatics express comic outrage at the fact of being compared to an American band, The Replacements, and being denigrated as having less talent than the Beatles or the Byrds. Rather than selling their souls to the music industry, whose power and control is metaphorically represented by Graceland mansion, the band emphasizes that it will stay true to its Canadian roots, and symbolically desecrate and destroy the heart of commercial American rock music—the resting place of Elvis Presley. Here, Martin Tielli and Dave Bidini, who authored this song, especially critique American commercialism in the music industry, and its effects on people. Despite (or perhaps because of) his success and appreciation by fans, the constant media attention and the push to produce more drove Elvis Presley to his own self-destruction (Brown and Broeske 1998; Guralnik 1999). Thus, he embodies the quintessential evil of American commercialism, and symbolically functions as the "other" to the far less commercial, "not-American" brand of music played by the Rheostatics.

Peace, order and good noise: Public space, homelessness and the conservative reforms of the Harris Government, 1995-2002

In addition to being vocally critical of what they perceive as the influence of American commercialism on Canadian society in general, the Rheostatics have been outspoken opponents of social reform in Canada and in their home province of Ontario, and were actively involved in protesting the "common sense" welfare reforms of the mid-1990s that were implemented by the conservative government of Ontario Premier Michael Harris, who led the province from 1995 to 2002 (Montgomery 2002). These reforms fundamentally reshaped the understanding of regional politics in Canada at the time, and openly dismantled the extensive welfare systems that citizens of this province had come to rely on. Furthermore, while advocated as getting hundreds of thousands of people off the welfare roles, it led to increasing rates of homelessness in Toronto and other large cities in Ontario, especially among the younger population (Mallan 2002). Subsequently, the public presence of the so-called "squeegee kids," who would wash windshields at major intersections in Toronto, and ask for money in return, created a major debate about loitering in the city, and led to their removal from public space by an ordinance that criminalized panhandling (Keil 2002). The widespread regulation of public space through city ordinances and state laws, just as we find it in many US cities, led to increasing social tensions in Toronto, and lasted through the entire Harris administration (Mitchell 2003; Parnaby 2003). The conservative ransacking of public space, the situation of the homeless, and the conservative shift in Canadian politics are brilliantly portrayed in the songs "Bad Time to be Poor" and "These Days are Good for the Canadian Youth Party Alliance". Shortly after the Progressive Conservative Party won the Ontario provincial elections in 1995 and Mike Harris became premier, the Rheostatics started writing and recording the songs for the album *The Blue Hysteria* which was released in 1996. One of the most successful songs of this album was "Bad Time to Be Poor," a catchy, upbeat pop tune that cynically decries the increasing poverty in the province of Ontario:

It is a bad time to be poor. 'Cause we don't give a shit no more. If you want to go for help, don't look next door. The line's been drawn and staked outside. I see to trying to lay the blame. On the folks in charge who hide in shame. For growing up with an open purse. And learning not about being alive. *Haven't I done enough to burn out? Haven't I been there to help out?* It is a bad time to be young. What's left to us can't be undone. Without it riding on our backs. When young and poor go hand in hand. It is a bad time to be poor. And feeling winter through a crack in the door (Rheostatics 1996).

This song clearly addresses several of the issues raised above. First, the Rheostatics emphasize that at the time of the writing of this song, not only is "it a bad time to be poor" (i.e. poverty is becoming more prominent in Ontario), but that we also "don't give a shit no more" (i.e. widespread poverty is coupled with growing apathy by the general populace). The band identifies the culprits of this situation as the progressive conservative government of the province whose members it accuses of having grown up with "open purses" (i.e. plenty of disposable income) and having never learned "to be alive" (i.e. they are too concerned with appearance and are somewhat "stuck up"). Most importantly, they state that youth poverty and homelessness are interrelated, by emphasizing that "young and poor go hand in hand." Furthermore, they address a major issue with homelessness in Canada; the exposure to cold temperatures that is dangerous, but pose an unavoidable risk for the homeless. For them, winter can be literally and figuratively felt through a crack in the door. They may find themselves in a primitive, unheated shelter, where cold temperatures enter particularly at nighttime, or constantly face the prospect of being expelled from heated environments, when a crack in the door is opened to them to escort them out of a store, mall, or public building. Five years after this initial complaint about the conservative turn in Ontario politics, and after observing such socio-spatial consequences of the Harris government, the Rheostatics took another aim at right-wing politics in Canada by parodying an imaginary conservative youth organization in the song "These days are good for the Canadian Conservative Youth Party Alliance." Released as the opening track for their 2001 album *Night of the Shooting Stars*, the song does not focus on conservative politics in Ontario alone, but relates this political persuasion to the undisputed heartland of Canadian conservative politics—the province of Alberta:

Days are good for the Canadian Conservative Youth Party Alliance. These days are good for us now. Days are good for the C.C.Y.P.A. These days are good for us now. The sun comes up on the flat Edmonton street. Oh, this morning will be good for us! There will be no nonsense for the workers to sweep in the morning. Will be good for us! I chipped my eye tooth on the back of a urinal. I will snip your life with the skill of a tailor. Man. There has been a shift in the public's expectations. These days are good for us now. There's a Lenny in my Kravitz that must be removed. These days are good for us now. All together! Oh, this day will be good for us. Oh, pitiful every man. How I've learned to shake your hand. C.C.Y.P.A.C.C.Y.P.A.C.C. This is a difficult world, but there are opportunities here. Oh, these days could be good for you. There are programs in place that will cure you, that will cure you. Oh, these days could be good for you. When the sun goes down on the flat Edmonton streets. You will seek normal pleasures. There are sports teams with cheerleaders who will double as hookers. I chipped my eye tooth on the back of a urinal. You've got your pain and I've got my pain. But I don't complain. No I don't complain. I first snip your life with the precision of a tailor. Oh pitiful every man, how I've learned to shake your hand. Oh simple every man, how I've learned ... to shake your calloused hand (Rheostatics 2001).

Here, the Rheostatics locate the heart of Canadian conservative politics and its acceptance by more and more young Canadians in the province of Alberta. While the city of Edmonton itself is known for its more progressive population, and often affectionately referred to as "Redmonton," it is formally the seat of government and capital of the province of Alberta. For more than 30 years, the province has been ruled by consecutive conservative governments, and it is generally regarded as Canada's most individualistic, free-market, big-business friendly and socially conservative region. In the song, the sun metaphorically rises in Edmonton to signal a new era of conservative politics, decisively influenced by the imaginary youth organization mentioned in this song. The song is written from the perspective of a seemingly drunk, dispassionate young member of a conservative youth organization who (after "having his eyetooth chipped on the back of a urinal") talks down to an "every man," the average citizen, and reflects on the growing acceptance of conservative politics all over Canada. Such politics include, as mentioned here, a no-nonsense, "common sense" approach to welfare reform that removes many social services from the poor. In typically conservative fashion, the reduced access to services for the poor is justified as necessary, and having the same effect on every citizen (e.g. "You've got your pain and I've got my pain. But I don't complain"). True to conservative doctrine, the youngster in this text promotes self-reliance (e.g. "This is a difficult world, but there are opportunities here"), and apathy for many worker's need of social assistance (e.g. "I first snip your life with the precision of a tailor. Oh pitiful every man, how I've learned to shake your hand. Oh simple every man, how I've learned ... to shake your calloused hand"). In the final analysis, this song parodies the often hypocritical demands of conservative politicians who, as recent scandals have shown repeatedly, often cannot personally conform to their own high moral, ethical, or social standards. Furthermore, it situates these politics in a defined spatial context, identifying the heartland of conservative politics in Canada with the seat of government in the province of Alberta. Using a protagonist in this song who walks the streets of Edmonton bragging about "cheerleaders that double as hookers," the Rheostatics argue that the conservative policies that originated in the Edmonton of Ralph Klein's government are hypocritical and seek to mislead the working class. Symbolized by the dismay that the protagonist of the song displays when he mentions how he has learned to shake the calloused hand of a worker, the band highlight how politicians often pretend to support the entire population of a country, while cutting deals that favor the interests of big business. Former Ontario premier Mike Harris, for example, whom the band so despises in the songs above, was caught working for Hydro One, a power company, who indirectly paid him for lobbying the province of Quebec to support a transmission line linking the two provinces (Findlay 2004). Thus, while Alberta is epitomized by the Rhesotatics as the quintessential birthplace and heartland of conservatism, it is Mike Harris whose social and spatial impact on the province of Ontario the band decries repeatedly.

The Group of Seven, nature, and nationality in music of the Rheostatics

Geographers interested in the social construction of national identity have long been interested in the multiple ways in which collective identity is constructed with the help of metaphors of nature (Lehr 1983; Malkki 1992; Moore 2003). Hall (1994) conceptualized nations not just as political entities, but also as systems of cultural representations through which communities construct and reflect the ideas of collective identity. He treated national cultures and national identities as discourses—social practices that produce exemplary texts that people use to understand themselves and the social world around them and that affect and guide our actions. Subsequently, he argued that national cultures (re)produce cultural and national identity through socially conditioned language. National cultures construct collective identities by producing meanings that we can identify with. Those meanings, then, are included in stories that are told about the nation, represented by memories that connect a nation's present with its past, and communicated through certain idea(l)s and myths that project the nation into the future. Discourses of the nation thus reveal a certain ideological basis, as well as opaque forms of power and political control that construct the nation in distinctly gendered, raced and sexed representations through the selected use of language. Memories, narratives and myths about nations are communicated through speech acts and mean nothing without an attachment to place (Kuhlke 2004). In addition, Nast (1994, 61) argues that linguistic acts are inherently spatial and subsequently, we need to consider the "great spatial and material diversity of representational/ speech forms."

Here, I trace out the discursive construction of Canadian national identity and its representation in music lyrics, soundscapes, and performative acts by the Rheostatics, to show that representations of nature figure prominently in their distinctly rugged representation of Canadian national identity. In particular, I argue that the national identity constructed by the Rheostatics in their music is very much influenced by the rise and dominance of naturalistic nationalism in Canada, as it has evolved since the mid-nineteenth century. Kaufmann (1998, 685) argued that around 1850, "English Canadians still expressed a traditional, agrarian/biblical fear of wilderness depravity." Influenced by British romanticism and authors such as Sir Walter Scott and James Fenimore Cooper, Canadian novels were increasingly focusing on heroic figures who encounter native populations, experience living in wilderness, and emerge as emotionally strengthened, wholesome, and healthy characters. As increased immigration and industrialism led to rapid urbanization in Canada, the associated effects of pollution, overcrowding and social tensions inspired a wilderness movement that glorified the return to nature. Regular retreats beyond the boundaries of organized civilization were to have a rejuvenating effect on the emotional and social health of people, thereby strengthening their resolve (Braun 2002; Nash 1967). Once Canada became a nation in 1867, several politicians—most notably W.A. Foster and his Canada First party, argued for the development of a "national sentiment." The literature and visual arts movement of the time followed the political nationalism of the post-confederation period in the 1870s with decisively nationalistic adventure novels and landscape painting that portrayed the lives of Canadians in isolated locations and the rugged scenery of the northern frontier. By the early twentieth century, Roger Stanley Weir had composed the song "O Canada" that would much later become the official national anthem. Most symbolic for this period in Canadian history, the song described the country as the "true north strong and free," thus associating national character with its northern location and climate. Throughout the 1910s and 1920s, such emphasis on identifying Canada through simple geographical elements and its northern landscapes became much more prominent. Perhaps most influential in this regard was a number of landscape painters who were collectively referred to as the Group of Seven. In 1920, J.E.H. McDonald, Lawren Harris, A.Y Jackson, Arthur Lismer, Franklin Carmichael, F.H. Varley and Franklin Johnston formed the group with the intention to popularize and regularly display Canadian landscape art. Kaufmann (1998, 691) describes the impact of the Group of Seven and its impact on the representation of Canadian national identity as follows:

> After 1920 the group came together as a unit in what some view as a political act inspired by the cultural nationalism of the period. For example, the group spent a large amount of time writing and speaking to the public as a means of proselytizing their work. Group members also set out to paint the *rougher, rawer elements of the Canadian north (primarily the Shield country of Northern Ontario)* [emphasis added] in vivid, sublime strokes. In doing so, they quickly incurred the ire of the genteel, Imperial Canadian art establishment. Nevertheless, the group used this conflict to symbolize the tension between Canadian and British identity and became active propagandists of the cause of an independent, Canadian cultural nationalism.

Canadian art, as defined by the Group of Seven, sought to set itself apart from British aesthetics by highlighting elements of nature over the portrayal of social environments, to cultivate an independent cultural nationalism. Representations of a distinctly Canadian identity, as fostered by these artists, focused on the impact of northern climate on human sentiment: the artists celebrated the notions of a tough, rugged, clean, healthy and forested northern land that would instill a certain hardiness and mental as well as physical healthiness in people:

> For instance, one member commented that in the minds of the group, Canada was a long thin strip of civilization on the southern fringe of a vast expanse of immensely varied, virgin land reaching into the remote north. Our whole country is cleansed by the pristine and replenishing air which sweeps out of that great hinterland (Berger 1966, 21).

Such stereotypical representation of naturalized nationalism through mythologized landscapes is reproduced in many of the Rheostatics' songs and writing (Bidini 2001). In the song "When Winter Comes" from the album *Melville*, the band celebrates the cleansing, strengthening power of winter:

> In the blue Canadian winter, I'll follow your trail till your love becomes a snowbank hardened by gale. When ice appears on matchsticks and the salt trucks fail, and the coalmen hibernate through their alarms. In the blue Canadian winter an iceman roams. Building railroads made of iron, sweat and skin. When you become thawed out your love will swamp the tracks, and my heart will be restored with virgin blood (Rheostatics 1991).

The emphasis on the toughness of winter and the restorative power of spring that shape the character of individuals and relationships is complemented by the Rheostatics' glorification of the northern forests, the open expanse of this wilderness, the prairies and frontier spirit that was cultivated in these regions. In "Northern Wish," from the album *Melville*, the Rheostatics adoringly sing about the forest of northern Ontario as a "parliament of trees," where "we don't need mathematics and we don't need submarines to tell how far the land does go," thus emphasizing the vastness of this natural landscape (Rheostatics 1991). Following such emotive attachment to the regional landscapes of northern Ontario, the Rheostatics point out the significance of this region for the entire country, which is to wake us up from "our deep provincial eyes," and highlight that this is a quintessentially Canadian landscape of national significance. As Brown (1996) has suggested, there is "something in the Canadian temperament is seduced by this imposing wilderness, we don't feel lost in it. Instead, our landscape grounds us."

Given these strongly Canadian place references, it is perhaps not surprising that in 1995, the band was commissioned by the National Gallery of Canada to create a musical tribute to the Group of Seven. The resultant album *Music Inspired by the Group of Seven* was a recorded live performance of the group in the gallery's theater in Ottawa, accompanied by several additional musicians. During the recording, projected images of paintings by the Group of Seven were shown along with era film footage and audio clips. In order for the paintings to speak to the audience and "do the talking," almost the entire recording is instrumental, with only sporadic words uttered, and one full piece from a previous album dominating as the theme song—"Northern Wish," from the *Melville* album discussed above. This rough, edgy song that celebrates northern Ontario forests and links this landscape to the rest of Canada, from the "prairie spine" of Alberta "until it hits the shore," exactly resonates the goal of the painters by "evoking the melting landscapes and thrilling skies of the Group's cavalier artists and their nationalist movement" (Rheostatics 1996 [quoted from song lyrics and liner notes of the album]). Just as the Group of Seven was breaking paths beginning in the 1920s, and established a distinctly nationalistic style of artwork, so do the Rheostatics today embrace the label of being nationalistic ambassadors of Canadian music:

> The musical landscape in this country is changing fast. Like the *Group of Seven* before them, the band finds itself hitting stride and leading a movement that is laying down a definable Canadian sound. In Canadian music, like in the arts scene of the 1920s, that sound has to do more with a new confidence than anything else … *A lot of Canadian bands are coming to terms with being comfortable writing about Canada and where they come from, about putting their story in Canadian style? I think there is one in music, it inevitable, it's very subtle, but it's there. It exists, and you can't change it* [interview excerpts with Dave Bidini and Martin are emphasized above] (Brown 1996).

Conclusion

As a form of artistic communication, music is often an ephemeral medium. Bands come and go, song are written, played and forgotten, and only a few voices manage to survive in the business long enough to make a meaningful impact on larger populations. Those that do often evoke distinctive place imagery that listeners can identify with. One need only think of Lynyrd Skynyrd's "Sweet Home Alabama"— a song that exudes place attachment and a sense of loyalty to locality. This holds true for Canadian music as well. The Tragically Hip defined for Canadians that "the Great Plains" literally and figuratively begin "at the hundredth meridian" and reminded us that Thompson, Manitoba is located at 55 degrees north, a "latitude that weakens our knees." Along with them, the Rheostatics are arguably the second most prominent band that openly expresses their love of country and their intense national pride in the soundscapes and lyrics of their music. For over twenty years and in twelve albums, they created a unique sound that was distinctly their own. Their edgy, often noisy, yet always melodic tunes permeated and influenced the Canadian alternative rock scene like no other music; and their anti-American, anti-establishment, and progressive politics often celebrated the multicultural, multi-lingual country that they so much enjoy and envision. So it is perhaps intentionally symbolic that the last album of the band was entitled *2067*. While the Rheostatics ended their journey in 2007, the country will live on and celebrate its bicentennial sixty years later. In a symbolic move, the band is again ahead of its time. Even more becoming, one of the songs on this album imagined a country that will be celebrated by the very nature of its past. In *Polar Bears and Trees* the band again invokes the intricate relationship between nature and nation, to construct what will then be a 200-year-old Canadian nation, and they do by highlighting the geography, topography, and climate that helped shape it in the first place:

> In a land of nothing but polar bears and trees, the inlet and the drumlin, the lichen and the weeds, the mighty beaver building, the otter and the loon, an eagle on the mountain, dives and kills and eats a rabid coon. Hear the roaring silence. You live the missing thrill, the topographical silence, the boundless northern will (Rheostatics 2004).

References

Anderson, B.R. (1991), *Imagined Communities: Reflections on the Origins and Spread of Nationalism* (London and New York: Verso).

Anderson, C. (2007), Canada4Life. <http://www.canada4life.ca>.

Ayres, J.M. (1995), "National No More: English Canada," *American Review of Canadian Studies* 25:2-3, 181-201.

Beaudreau, S. (2002), "The Changing Face of Canada: Images of Canada in National Geographic," *American Review of Canadian Studies* 32:4, 517-46.

Berger, C. (1966), "The True North Strong and Free," in P. Russell (ed.).

Bernard, A. (1978), *What does Quebec Want?* (Toronto: James Lorimer & Company).

Bhabha, H. (1993), *Nation and Narration* (London: Routledge).

Bidini, D. (2001), *Tropic of Hockey: My Search for the Game in Unlikely Places* (Toronto: McClelland & Stewart).

Braun, B.P. (2002), *The Intemperate Rain Forest: Nature, Culture and Power on Canada's West Coast* (Minneapolis: University of Minnesota Press).

Brown, L. (1996), *Rheostatics* collaborate with Group of Seven. *Canadian Broadcasting Corporation* Television Program. 24 October 1996. <http://archives.cbc.ca/IDC-1-68-754-4635/arts_entertainment/group_of_seven/clip11>, accessed 17 September 2007.

Brown, P.H. and Broeske, P.H. (1998), *Down at the End of Lonely Street: The Life and Death of Elvis Presley* (New York: Signet).

Buckstein, S. (2005), "Socialized Medicine: A Symbol of Canadian National Identity?" *Capitalism Magazine*, 15 August.

CensusScope (2007), <http://www.censusscope.org/us/chart_language.html>, accessed 19 June 2007.

Clark, H.D. and Kornberg, A. (1996), "Choosing Canada? The 1995 Quebec Sovereignty Referendum," *PS: Politics and Society* 29:4, 676-82.

Conway, J. (2004), *Debts to Pay: The Future of Federalism in Canada* (Toronto: James Lorimer & Company).

Desbiens, C. (2004), "Producing North and South: A Political Geography of Hydro-Development in Quebec," *Canadian Geographer* 48:2, 101-18.

Doucet, M.J. (1998), "A City 'Waiting for the Sunrise': Toronto in Song and Sound," *Canadian Journal for Traditional Music* 26 (Online edition without numbered pages, <http://cjtm.icaap.org/contentt/26>).

Duffy, M. (2005), "Performing Identity within a Multicultural Framework," *Social and Cultural Geography* 6:5, 677-92.

Dyck, R. (2000), *Canadian Politics: Critical Approaches* (Scarborough: Nelson Thomson Learning).

Findlay, A. (2004), "Harris Tipped 18.5 G Foe Hydro One Advice," *The Toronto Sun*, 18 March, p. 10.

Gee, J.P. (2005), *An Introduction to Discourse Analysis: Theory and Method* (London: Routledge).

Grant, G.P. (2005), *Lament for a Nation: The Defeat of Canadian Nationalism* (Montreal: McGill-Queen's University Press).

Guralnick, P. (1999), *Careless Love: The Unmaking of Elvis Presley* (Boston, New York, London: Little, Brown and Co.).

Halbwachs, M. (1985), *Das kollektive Gedächtnis* (Frankfurt: Suhrkamp).

Hall, S. (1994), *Rassismus und kulturelle Identität. Ausgewählte Schriften* (Hamburg/Berlin).

Hill, C. (1995), *The Group of Seven: Art for a Nation* (Ottawa: National Gallery of Canada/McClelland and Stewart).

Ho, C. (2000), "Popular Culture and the Aestheticization of Politics: Hegemonic Struggle and Postcolonial Nationalism in the Trinidad Carnival," *Transforming Anthropology* 9:1, 3-18.

Hunt, K. (2007), "Static Electricity: The Rheostatics' Great Farewell," *Torontoist,* 29 March. <http://www.torontoist.com/archives/2007/03/static_electric.php>.

Jackson, D.J. (2005), "Peace, Order and Good Songs: Popular Music and English-Canadian Culture," *American Review of Canadian Studies* 35:1, 25-4.

Jacobson, N. (2004), "Before You Flee to Canada, Can We Talk?" *The Washington Post*, 28 November, p. B2.

Kaplan, D.H. (1994a), "Population and Politics in a Plural Society: The Changing Geography of Canada's Linguistic Groups," *Annals of the Association of American Geographers* 84:1, 46-67.

Kaplan, D.H. (1994b), "Two Nations in Search of a State: Canada's Ambivalent Spatial Identities," *Annals of the Association of American Geographers* 84:4, 585-606.

Kaufmann, E. (1998), "Naturalizing the Nation: The Rise of Naturalistic Nationalism in the United States and Canada," *Comparative Studies in Society and History* 40:4, 666-95.

Keil, R. (2002), "'Common Sense' Neoliberalism: Progressive Conservative Urbanism in Toronto, Canada," *Antipode* 34:3, 578-601.

Kuhlke, O. (2004), *Representing German Identity in the New Berlin Republic: Body, Nation and Space* (Lewiston, NY; Queenboro, ON and Lampeter, UK: Edwin Mellen Press).

Lecker, R. (1993), "'A Quest for the Peaceable Kingdom': The Narrative in Northrop Frye's Conclusion to the *Literary History of Canada*," *PMLA* 108:2, 283-93.

Lees, L. (2004), "Urban Geography: Discourse Analysis and Urban Research. *Progress in Human Geography* 28:1, 101-107.

Lehr, J.C. (1983), "'Texas (When I Die)': National Identity and Images of Place in Canadian Country Music Broadcasts," *Canadian Geographer* 27:4, 361-70.

Linke, U. (1999), *German Bodies* (London, New York: Routledge).

Lipset, S.M. (1991), *Continental Divide: The Values and Institutions of the United States and Canada* (New York and London: Routledge).

Mallan, C. (2002), "The Legacy of Mike Harris," *Toronto Star*, 16 March, p. H2.

Malkki, L. (1992), "National Geographic: The Rooting of Peoples and the Territorialization of National Identity Among Scholar and Refugees," *Cultural Anthropology* 7:1, 24-44.

Millard, G., Riegel, S. and Wright, J. (2002), "Here's Where We Get Canadian: English-Canadian Nationalism and Popular Culture." *American Review of Canadian Studies* 32:1, 11-34.

Mitchell, D. (2003), *The Right to the City: Social Justice and the Fight for Public Space* (New York: Guilford Press).

Molson Brewery Inc. (2000), *I am Canadian*. Television Advertisement.

Montgomery, B. (2002), *The Common (Non)sense Revolution: The Decline of Progress and Democracy in Ontario* (New Providence: BPR Publishers).

Moore, D.S. (2003), *Race, Nature and the Politics of Difference* (Durham, NC: Duke University Press).

Nash, R.F. (1967), *Wilderness and the American Mind* (New Haven, CT: Yale University Press).

Nast, H. (1994), "Women in the Field: Critical Feminist Methodologies and Theoretical Perspectives," *The Professional Geographer* 46:1, 54-66.

O'Connor, A. (2002), "Local Scenes and Dangerous Crossroads: Punk Theories of Cultural Hybridity," *Popular Music* 21:2, 225-36.

Obert, J.C. (2006), "The Cultural Capital of Sound: Quebecite's Acoustic Hybridity." *Postcolonial Text* 2:4, 2-14.

Parnaby, P. (2003), "Disaster through Dirty Windshields: Law, Order and Toronto's Squeegee Kids," *Canadian Journal of Sociology* 28:3, 281-307.

Rayner, B. (2007), "*Rheostatics*' Swan Song," *Toronto Star*, 29 March. <http://www.thestar.com/article/196682>.

Reich, R. (1999), "Good and Bad Nationalism," *The American Prospect*, 29 November. <http://www.prospect.org/webfeatures/1999/11/reich-r-11-29.html>.

Russell, P. (ed.) (1966), *Nationalism in Canada* (Toronto: McGraw Hill).

Sachdev, I., Arnold, D.Y. and Yapita, J.D. (2006), "Indigenous Identity and Language: Some Considerations from Bolivia and Canada," *Birkbeck Studies in Applied Linguistics* 1, 107-28.

Salloum, H. (2004), "The Two Faces of Tourism: Why Do Canadians Travel Abroad?" *Contemporary Review*, March, 37-47.

Small, C. (2000), "Nationalism and Difference in the Cosmopolitan City of Montreal," *Geography Research Forum* 20, 70-85.

Smith, A. (1991), *National Identity* (Reno: University of Nevada Press).

Smith, S.J. (1997), "Beyond Geography's Visible Worlds: A Cultural Politics of Music," *Progress in Human Geography* 21:4, 502-29.

Statistics Canada. (2001), Proportion of immigrants born in Europe and Asia per period of immigration, Canada, 2001. <http://www12.statcan.ca/english/census01/products/analytic/companion/etoimm/canada.cfm>, accessed 6 June 2007.

Stevenson, G. (2004), *Unfulfilled Union: Canadian Federalism and National Unity* (Montreal: McGill-Queen's University Press).

Thomas, D. (ed.) (2000), *Canada and the United States: Differences that Count* (Toronto: Broadview).

Waiser, B. (1995), *Park Prisoners: The Untold Story of Western Canada's National Parks, 1915-1946* (Saskatoon: Fifth House Publishers).

Ward, P. (2003), *White Canada Forever: Popular Attitudes and Public Policy Toward Orientals in British Columbia* (Montreal: McGill-Queen's University Press).

Wilson, E. (2005), "Gender, Nationalism, Citizenship, and Nunavut's 'Territorial House': A Case Study of the Gender Parity Proposal Debate," *Arctic Anthropology* 42:2, 82-94.

Wright, R. (2004), *Virtual Sovereignty: Nationalism, Culture and the Canadian Question* (Toronto: Canadian Scholar's Press).

Yalnizyan, A. (2000), *Canada's Great Divide: The Politics of the Growing Gap Between Rich and Poor in the 1990s* (Toronto: Center for Social Justice).

Young, R.A. (1994), "The Political Economy of Secession: The Case of Quebec," *Constitutional Political Economy* 5:2, 221-45.

Young, R.A. (1998), *The Secession of Quebec and the Future of Canada* (Montreal: McGill-Queen's University Press).

Discography

lang, kd. (2004), *Hymns of the 49th Parallel.* Compact Disc. Nonesuch Records.

Rheostatics (1991), *Melville.* Compact Disc/LP. Green Spouts Music Club/ DROG/Intrepid.

Rheostatics (1992), *Whale Music.* Compact Disc/LP. Sire/Warner Brothers.

Rheostatics (1996), *Music Inspired by the Group of Seven.* Compact Disc/LP. Green Spouts Music Club/ DROG/Intrepid.

Rheostatics (1996), *The Blue Hysteria*. Compact Disc. Cargo Records.

Rheostatics (2001), *Night of the Shooting Stars*. Compact Disc. Perimeter Records.

Rheostatics (2004), *2067*. Compact Disc. True North Records.

Part V
Local Music in a Connected World

The next three chapters address distinct patterns of local production and consumption of music and how globalizing forces have changed, although not always fundamentally altered, such patterns. As noted in the introduction to Robert Kruse's chapter on John Lennon and Yoko Ono, media are an emerging topic in geography. In the chapter "Internet Radio and Cultural Connections: A Case Study of the St John's, Newfoundland Radio Market," Sara Beth Keough emphasizes not only media, but also themes that are found elsewhere in this book: diasporic communities, music as a vessel for place identity, technological change, and the politics of Canada as seen and heard through the lens of music. At the heart of the argument in Keough's article is that new technology can have a dramatic impact on the social meaning of music. In her case, Newfoundland's diaspora, which is found throughout North America due to the island's poor economic development, has a greater opportunity today than in the past to access music from their homeland. The habit of listening to Internet radio broadcasts from your homeland is an increasingly global phenomenon, and here we gain further insight into this trend. Newfoundland's geographic isolation and its twentieth century history as a relatively recent addition to the Canadian state have resulted in a strong sense of regionalism. Unlike the Rheostatics, as discussed earlier by Kuhlke, Newfoundland music does not represent "Canadianness" but rather a distinctly regional identity. The music from the province may not have developed a strong ethos of nationhood; instead, it generally focus on the everyday life on the island. As such, the emotive quality for the diaspora is often one of nostalgia for home. Nor does Newfoundland music necessarily exhibit musical cohesiveness as a "scene," although the Scot-Irish heritage of Newfoundland is commonly heard in the music.

Many radio stations in Newfoundland dedicate some airtime to local music. As radio has changed from a medium that was confined by the geographic reach of the airwave signal to becoming truly global with the advent of the Internet, diasporic communities have a new way of connecting with their homelands. Radio content everywhere is dominated by music, so logically, music occupies a central position as a transmitter of culture in a way that perhaps was not the case in the past. Although the music itself as well as the geographical location differ, Keough's argument is similar to the subsequent chapter by Holly Kruse. Both have identified significant changes in how music is produced and consumed with the advent of the Internet. New technologies have always changed music, as well as the social spaces in which music is played and listed too. The electrification of music and the development of recorded music—the phonograph—had such effects in the past. Now, the Internet is transforming some aspects of how music is received.

Holly Kruse is one of the keenest observers of local music scenes in the United States. She has followed "indie" rock in Urbana-Champaign, Illinois and other places since the early 1990s (1993, 2003). Much like Robert Kruse's contribution, this chapter represents an extension of an existing body of work—in Holly's case on alternative music scenes and in Robert's case on the Beatles. They are both clearly authorities in their respective field within music geography (although Holly Kruse's academic field is Communications, her research is eminently spatial in orientation). In "Local Independent Music Scenes and the Implication of the Internet" she updates her previous research to include the impact of new telecommunications technology on the functioning of local scenes. In essence, Kruse's chapter is a description of "before and after" the Internet, making a general argument rather than offering a specific case study. Exactly how the Internet impacts existing social, economic, and cultural geographic patterns have been subject to much debate. Local music scenes, which are based on networks comprised of individuals who communicate with each other largely on a face-to-face basis, are phenomena that intuitively must be altered with the introduction of a new distance-annihilating technology such as the Internet. Kruse shows, however, that the Internet does not produce "the end of geography" in the case of local music scenes, which are still thriving and relevant. She concludes that local identity and local sounds are still evident despite growing translocal connections fueled by the Internet. The local and the global can operate in concert, rather than being mutually exclusive. The "virtual" and the "real" are not entirely separate spheres. On a macro level, Ola Johansson and Thomas Bell similarly suggest in Chapter 13 that the broad geography of alternative rock in North America has not undergone dramatic changes in an era of globalizing culture and technology.

One obvious aspect of the Internet is how the amount of available information has increased. Today, music fans can turn to Internet radio stations for outside-the-mainstream music that otherwise may be marginally available in many locations. Numerous other popular formats increase the access to music, such as YouTube, where videos and concert performances are available to areas with poor performance and video viewing opportunities. Concrete and conventional spaces that occupied central positions in the network of local scenes may, on the other hand, be declining. The local college radio station is exposed to more competition, as is the record store, which, in the past, functioned as a meeting place as much as a retail location. The meeting place function is much more characteristic for indie scenes (and other music subcultures) than for the record purchasing general public, so the shift to Internet-based music commerce may have a relatively smaller impact on local music scenes of the character described by Kruse. Moreover, on an ideological level, the locally grounded tradition of music scenes still serves, perhaps even more so in the Internet-age, as a guarantee that the music is "real" and that it emerged from the ground up rather than being manufactured for the greatest possible audience and profit. Local connections implies "authenticity"—a concept which is not unique to indie rock, but arguably more important here than in many other musical genres as indie by definition implies a certain psychological and aural distance from mainstream music. Place therefore exists in opposition to placelessness, and takes on an ideological meaning.

Another example how the Internet does not "change everything" is the concert—music's version of face-to-face meetings in the business world where proximity is still deemed necessary. The concert experience today is one major element of the music industry that has not experienced a decline (Black, Fox, and Kochanowski 2007). In local indie scenes, concerts are also still important. The Internet, despite its global connectedness, can actually improve local marketing and information dissemination about performances. This shows how the technology can improve local network ties. Not everyone is necessarily well connected to cyberspace, but because of their typical demographic profile indie fans are mostly an Internet savvy group.

Kruse is also, together with Ola Johansson and Thomas Bell's chapter "Where are the New US Music Scenes?" bringing indie/alternative rock and locality into focus. Local rock scenes, as we know them, had a breakthrough in the 1980s post-punk era where college radio, the emergence of new labels, and several other factors brought the music to prominence.

Today, local scenes are not as prominent in the popular imagination. Local scenes lost their allure, yet Johansson and Bell argue that it is in places well known for alternative rock where new music is most likely to emerge even today. In order to properly understand place-based creativity in new American rock music, Johansson and Bell have mapped the place of origin of new rock bands. The data, which are primarily based on college radio plays, reveal the spatial characteristics of alternative rock and its several subgenres. More than a decade has passed since grunge defined the Seattle scene (Bell 1998) and, before that, REM and B-52s gave Athens, Georgia cachet for alternative rock. But to write about Seattle and similar places as scenes no longer makes for "sexy" copy in the music press. The authors argue, therefore, that we must pay attention to practices, particularly in media, that frame the emergence and departure of scenes. The discursive element to scenes has been recognized as a primary tool to connect places and music. Such a connection is generally made by media and music industry to authenticate music and to situate it within a particular genre. In the twenty-first century, while the authentication of music in a globalized world is still important, Johansson and Bell suggest that it is common to do so using translocal discourses as much as, and perhaps even more than, local scenes.

References

Bell, T.L. (1998), "Why Seattle? An Examination of an Alternative Rock Hearth," *Journal of Cultural Geography* 18:1, 35-47.

Black, G., Fox, M. and Kochanowski, P. (2007), "Concert Tour Success in North America: An Examination of the Top 100 Tours from 1997 to 2005," *Popular Music and Society,* 30:2, 149-72.

Kruse, H. (1993), "Subcultural Identity in Alternative Music Culture," *Popular Music* 12:1, 33-41.

—— (2003), *Site and Sound: Understanding Independent Music Scenes* (New York: Peter Lang).

Chapter 11

Internet Radio and Cultural Connections: A Case Study of the St. John's, Newfoundland Radio Market[1]

Sara Beth Keough

Introduction

Globally, radio remains the most important medium of mass-communication. According to communications and media studies scholar Andrew Crisell, "the obvious advantages of modes of mass communication are that the sender can communicate with multitudes of receivers at the same time and at distances beyond that achievable by inter-personal communication" (1994, 4). No longer is physical proximity a requirement for communication. Radio, as form of mass communication, functions through an auditory medium that relays words and voices. In turn, these auditory signals invoke our imagination and force us to create images in our mind relative to the words and voices heard.[2] Significant advances in radio technology have allowed listening to evolve from being an activity tied to a fixed location, to an activity that, in many cases, has become secondary to other tasks and can encompass a great amount of mobility (Berland 1994; Crisell 1994). For example, one might listen to the radio while driving a car, or while cleaning the kitchen. The number of activities an individual can pursue while listening to the radio has increased with changes in technology, as radios have become smaller and thus more portable.

1 The author would like to thank Dr. Thomas Bell for his unwavering support and work in advising this project, and also Scott Youngstedt for his insightful comments and suggestions on drafts of this chapter. This research was funded through grants from the Canadian Embassy, the Association of American Geographers, the Society for Women Geographers, the Cultural Geography Specialty Group, the McCroskey Foundation, and the W.K. McClure Foundation for the Study of World Affairs.

2 A recent *New York Times* article discussed the recent trend that radio stations are adding video content to their websites, which adds a visual element to what was once purely auditory (Siklos 2007). While further discussion on this new visual component is beyond the scope of this chapter, it is something that scholars of radio and media will need to address in the future.

The mobility of radio is not just specific to the listener's location. Advances in technology have made the radio signal mobile as well. The Internet, for example, has expanded the reach of a radio station's signal beyond that of the terrestrial signal. This signal expansion has, in turn, expanded the listener base of radio stations to include those well beyond the reaches of a station's terrestrial signal. Listener base expansion has, in many cases, been advantageous for radio stations. This study examines the impact of Internet broadcasting in the radio market in St. John's, Newfoundland, the provincial capital of the Atlantic Canadian province of Newfoundland and Labrador. In this chapter, I argue that the local music on St. John's radio stations and Internet broadcasts, which often contains Newfoundland references and stories, serves as a cultural connection point between Newfoundlanders living on the island and those that have moved off the island and are living "away."[3]

The St. John's radio market, and its eleven radio stations, was chosen as a case study for a number of reasons. The Avalon Peninsula, where the city of St. John's is located, is the home to two-thirds of the province's population, and about half of the island's population lives in the St. John's metropolitan area itself. With eleven radio stations, the St. John's radio market reaches more than half the province's population with its terrestrial signals and transmitters, and is the largest radio market in the province. Historically, Newfoundland played an important role in the evolution of communication technology; it was the final destination for the first underwater cable between Europe and North America in 1866. And, St. John's was the recipient of the first transatlantic wireless radio signal sent from Europe. Moreover, the St. John's radio market has two radio stations that have been broadcasting since the early 1920s. The continuous presence of radio broadcasting on the island has meant Newfoundlanders could rely on radio technology for news and entertainment. Radio listening thus became a tradition on the island.

Moreover, Newfoundlanders feel that their cultural identity is distinct from the rest of Canada. This feeling of distinctiveness is rooted in Newfoundland's history. Like much of North America, Newfoundland was once a colony of Great Britain, but, in 1855, the island and part of the mainland gained dominion status, and essentially functioned as an independent country. In 1934, after periods of economic decline, Newfoundland became the first territory in history to democratically choose to return power to its former colonizer. In 1949, however, Newfoundland's economy had recovered, and Newfoundland and Labrador became Canada's newest province after voting 51 percent in favor of confederation to 49 percent in favor of remaining an independent country. This narrow margin meant that many Newfoundlanders did not favor confederation and did not consider themselves to be Canadians, at least culturally. As government programs infiltrated the new province, Newfoundlanders were challenged to maintain unique aspects of

---

3   This term is the local parlance that carries with it the implication that those who leave will return some day.

their culture (accents and language, music, traditions, lifestyles) amidst possible assimilation to Canada culture.

In recent years, Newfoundlanders have made efforts to preserve this distinct cultural identity in light of the massive out-migration to mainland North America that the islanders have has experienced since 1992, the year the Canadian government banned cod fishing in the North Atlantic that suddenly put more than 30,000 Newfoundlanders out of work.[4] In its 2003 report, *Reviewing and Strengthening Our Place in Canada*, the Royal Commission in Newfoundland stated that out-migration was "the most significant social and economic challenge facing the province, especially in rural Newfoundland and Labrador" (35). Since 1992, a greater percentage of men than women have left the province (53.9 percent of the total in 2002). Most migrants are young adults and families with children. This trend, coupled with declining birthrates, has changed the age structure of the province as the average age of Newfoundland's population is rising. The Commission attributes this out-migration to Newfoundland's struggling economic situation and high unemployment rates—14.8 percent in 2006 (Statistics Canada 2007a).

High wages offered by the oil industry in Alberta make that province a popular destination for Newfoundland migrants.[5] In their study on how internal migrants in Canada use computer-mediated communications, Hiller and Franz (2004) found that a majority of the Internet communications traffic among migrants to Alberta was with Newfoundland. Interviews with inter-provincial migrants to Alberta revealed that one of the ways the migrants from Newfoundland used the Internet was to listen to Newfoundland radio stations, especially ones with a significant amount of Newfoundland musical content. The authors cited distance, a long tradition of out-migration (from Newfoundland), and a strong cultural identity in the homeland as reasons for this strong connection. Hiller and Franz's findings help to substantiate the importance that Newfoundland radio broadcasters and listeners place on radio as a connection mechanism between those living on the mainland and the island. So much of the content on St. John's radio stations relates to Newfoundland that the radio has become a means of cultural expression in both areas, more so than other forms of mass media communication, such as television. The St. John's radio market thus provides an interesting example of the link that music radio broadcasting on the Internet provides between Newfoundlanders living in diaspora and those who remain on the island.

---

4   Since 1992, Newfoundland has lost 12 percent of its population, or close to 70,000 people (Statistics Canada 2007b).

5   While Toronto has also been a popular destination for Newfoundland migrants, economic opportunities in Alberta as a result of the oil boom attract 48 percent of Newfoundland migrants (Statistics Canada 2007c).

Exploring radio's local and global attributes

The study of radio broadcasting in Newfoundland presents an example of "glocalization," a connection between local and global phenomena. On the one hand, because of new radio technologies such as Internet and satellite radio, what used to be a fairly local industry now has global reach. On the other hand, as this chapter shows, radio is being used to foster and reinforce a strong local identity, and thus connects the Newfoundland diaspora with its cultural hearth on the island.

Radio industry and globalization

Terrestrial radio signals are limited by topography and signal strength. A radio station's audience had to be within range of the station's signal or transmitter to receive broadcasts. Internet radio broadcasting has significantly changed the face of broadcasting, enabling stations to reach global audiences. Hendy (2000, 21) refers to radio as "the most local of the electronic mass-media, yet the first one to be distributed on a global scale." This local-global linkage that radio provides led geographer Peter Dicken to refer to the medium as the "most significant global media" (2003, 97-8). Because Internet radio broadcasting has the power to link global audiences to those listening via a station's terrestrial signal, radio has created a new "space of flows" (Castells 2000) where interaction between people is dynamic, rather than fixed, and can occur with or without physical proximity, capitalizing upon a system of interregional networks or nodes that the web of computers and Internet connections provides. As in the case of Newfoundland, this new space of flows created by Internet radio broadcasting means that Newfoundland radio has not only become more global in its reach, but at the same time focused more programming on local content, attracting a listening audience among the Newfoundland diaspora.

Radio and ideas of community and identity

Radio used to be the primary form of mass communication in North America, and it still is the primary form globally. Before the development of television, radio facilitated cohesion and interaction between those people already in proximity to each other. Stern (2003) notes that family togetherness was often created through radio, as it often occupied a central place in the home, and he paints a picture of the family gathered around the radio listening together (see also McGee 1985; Webb 1994). Stern shows that historically the radio was used in the cultural context of the family and for the specific cultural practice of providing evening entertainment in the home.

Radio's role in North American family life was challenged, and in many cases replaced, by the invention of television. The radio did not completely vanish from societal use, however, because it benefited from two technological innovations: the transistor, which made receivers smaller and cheaper, and the expansion of broadcasters to the FM spectrum, which enhanced signal strength and clarity of sound. These innovations made radio an ideal medium for broadcasting music, and thus radio changed its focus from broadcasting to the population at large to broadcasting for niche markets. The number and variety of radio stations grew as smaller stations sprang up to capture smaller, more narrowly-defined audiences (Grossberg et al. 1998).

Before television became commonplace, radio also played a role in establishing national identities, especially in the case of national radio stations, which were sometimes the only ones available. This happened not so much through content, but as a "national, temporal symbol" encouraging everyone to tune in at a particular time (Moores 1993). Radio acted as a "speaking clock" (Hay et al. 1996), which announced the time periodically during broadcasts. This type of control mechanism helped to synchronize the private activities of the listeners even when great distances separated them. The idea that everyone in the country or region listened to the same thing at the same time is an example of Anderson's (1983) "imagined community." Listeners did not meet, and in many cases did not interact, but because radio shows are often broadcast based on demand, listeners knew they were part of a larger listening community.

Radio listening as a cultural practice survives the wide dissemination of television because its role as a present-day form of technoculture continues to change. The changing role of technology has led to new ways of examining the interaction between culture and technology. Ethnomusicologist Deborah Wong states that technology itself is culture and that "an examination of technological practices in context is the only way to get at what technology 'does'" (Wong 2003, 125). Jeffrey Webb, a scholar of Newfoundland radio history, similarly argues that "relating culture to the material basis of society has shown how new technologies and new social structures are created or adapted so that the new technology fits into existing social relations" (Webb 1994, 5). In other words, those who adopt new technology (such as radio) are actively shaping its use and purpose to society. Thus, we must look not only at radio itself, but also at what radio conveys to listeners, and how it conveys its message, through the use of technology. Canadian radio scholar Jody Berland (1994, 176) argues that radio serves as a medium for "the enactment of a community's oral and musical history" because it is a means through which broadcasters, for example, convey something of meaning to listeners, who then decide if what they hear is meaningful. To fully appreciate the radio as a form of technoculture, we must examine both those who produce and consume this aspect of culture: the audience.

Internet radio in a larger Canadian context

To understand the broader scope of Internet radio broadcasting in Newfoundland, I will first discuss the evolution of Internet radio broadcasting policy in Canada. St. John's radio stations fall under the jurisdiction of the Canadian Radio-

Television and Telecommunications Commission (CRTC), a branch of the federal government that regulates broadcasting policy. Therefore, St. John's radio stations have to be cognizant of both their audience and the requirements set by the federal government.

By the early 1990s, the Canadian government recognized the necessity of connecting the country to the Internet. In 1994, the Canadian government created the Information Highway Advisory Council of Canada (IHAC). The Council "made suggestions for policies and programs relating to access, Canadian content and competitiveness on the Internet" (Turk and Johnston 1997, 1). The IHAC wanted an equal balance among wealth creation, social cohesion, and political liberty on the information super highway. Recommendations made by the IHAC were designed to create such a balance by assuring that Canadians had affordable access to the Internet, that Canadian content was present on the Internet, and that economic competitiveness on the Internet was sustainable. It became clear to the Council that new measurement techniques for Canadian content were needed for the Internet, and that these techniques would constantly evolve and change (Turk and Johnston 1997).

Historically, the public and private sectors in Canada have collaborated on many social and economic issues. The Internet is no exception.[6] Through collaborations such as CA.net, the Community Access Program (CAP), and the SchoolNet program, which networked all the schools in Canada, Internet use and access has grown in Canada. The Internet is exceptionally important in Canada because access to it helps bridge the expansive geographical divides among the nation's population (Turk and Johnston 1997). As of March 2008, over 84 percent of Canada's population had access to the Internet, according to the International Telecommunications Union.[7]

One of the key objectives of the IHAC was to maintain the cultural presence of Canada on radio and television as directed by the new Broadcast Act of 1991. IHAC stated that:

> Policies to promote Canada's cultural identity have never sought to protect Canadians from exposure to foreign cultural content. What Canada has traditionally sought to preserve—in the context of exceptionally open access to foreign cultural content—is a measure of control by Canadians over our own cultural markets and our ability to create, produce and make available our own cultural content (IHAC Phase II Report 1997, 57-8).

Ensuring Canadian content on the Internet presented new challenges. Canadian content regulations, which set minimum quotas for the amount of Canadian-

---

6   Under the Constitution Act of 1867, telecommunications fell under federal jurisdiction.

7   In the United States, according to Nielson/NetRatings, 72.5 percent of the US population has access to the Internet.

produced material in television, radio, and film, traditionally separated the audio (radio) from the visual (television) in their directives. The Internet, however, carries both audio and visual elements.

CRTC regulations also traditionally separated broadcasting and communication technologies.[8] The Internet combined these industries, and forced the IHAC and the CRTC to reconsider their definition of "programming." Regulating the Internet became a highly debated issue because Internet content was not just Canadian (Turk and Johnston 1997). Furthermore, the Internet was defined as a "private communications medium" where ideas are exchanged among individuals. The Canadian government does not regulate content of private communications. Thus, the advent of the Internet and its role as both a broadcaster and as a communications device further complicated the regulation debate (Canadian Content and Culture Working Group 1995). The IHAC recommended that the public and private sector continue their collaboration "to strengthen Canada's linguistic and cultural distinctiveness by encouraging Canadian Content providers to maximize their opportunities to employ the Internet for the delivery of content-intensive products and services" (Turk and Johnston 1997, 13-15). Measures used to encourage access to Canadian content on the Internet included the digitization of material in archives, museums, libraries, and government agencies, in addition to radio and television broadcasting via the Internet.

Most radio stations that offer "listen live" options via their website broadcasts offer the Internet listener exactly what they also air via terrestrial signals. In this sense, stations are maintaining a Canadian presence through Internet broadcasting. Prior to the Internet, a listener within the political boundaries of Canada could turn on the radio and choose between a number of stations, most of which played Canadian music to some degree.[9] With the Internet, however, a listener has the option to choose between listening to a Canadian radio station, where they will hear at least some Canadian music, or to a foreign station, which may or may not play any Canadian music at all. In this sense, the argument that Canadian content regulations restrict the information consumers receive is no longer valid (Stanbury 1998). Internet users in Canada who want to listen to the radio can choose a US station that plays exclusively music from the United States. The interesting question now is, what Internet listening choices are Canadians, and Newfoundlanders in particular, making? And, more importantly, who drives the content on the Internet—the producers of content, or the consumers of that content?

---

8   The broadcasting sector of the CRTC includes radio and television broadcasting. The communications sector includes telecommunications.

9   Prior to Internet radio broadcasting, Canadians had some access to US radio stations, none of which had Canadian Content requirements.

Methodology

Data for this study were collected between 2003 and 2006 in Newfoundland. Information was gathered from two groups of participants in the St. John's radio market: station personnel (i.e. DJs, program directors, station managers, music librarians for radio stations in the St. John's radio market) and listeners, both in rural and urban communities within the St. John's radio market. The inclusion of listeners from both rural and urban areas is important in understanding linkages that radio provides because in Newfoundland there are subcultural differences between the rural areas (called outports) and the urban area of St. John's. The St. John's radio market reaches both rural and urban listeners. To effectively study the audience of St. John's radio stations, I had to include listeners in both locales. Because Hiller and Franz's (2004) research address the Newfoundland diaspora, I decided to focus exclusively on radio listeners on the island.

   With both groups of participants, my questions used the term "Newfoundland music." As an etic researcher approaching the topic as an outsider, I let my participants define what they meant by Newfoundland music, and I let their definitions drive my research. Naturally, definitions varied among participants, but three ideas emerged that were common to most of the definitions. First, Newfoundland music was defined as music produced or performed in Newfoundland and/or by Newfoundlanders. This part of the definition is important because Newfoundlanders who did not live on the island but still wrote music, were considered to be producing Newfoundland music. So, physical proximity of the musicians to Newfoundland was not necessarily a requirement. Second, my participants defined Newfoundland music by its instrumentation. For example, many of my participants stated that if the music had an accordion, which is a popular instrument in Newfoundland and which connects music in Newfoundland to many of its Irish and English roots, then it is considered Newfoundland music. Finally, if the music and lyrics addresses Newfoundland as a place, Newfoundland culture and traditions, or social and economic issues in Newfoundland, it is considered by my participants to be Newfoundland music.

   Because much of the music from Newfoundland addresses social, political, and economic issues in Newfoundland culture, such as the hardships of the Cod Moratorium, or Newfoundlanders' connection to the ocean, music in Newfoundland culture has become the means through which Newfoundlanders have fostered a separate identity (Keough 2007). The stories told through Newfoundland music reflect history and traditions that Newfoundlanders have in common. Furthermore, music-making in Newfoundland is often a communal activity. "Kitchen parties," for example, are well-known, typically spontaneous, music-making events in Newfoundland culture. Friends and family gather in each other's houses, usually in the kitchen. Before the widespread use of electric heat, the kitchen was the warmest room in the house. Singing often ensues at these kitchen parties and instruments ranging from guitars and fiddles to spoons, pots and pans accompany the performers. The songs sung are usually ones that everyone in the group knows. Because much of the music produced in Newfoundland tells stories of Newfoundland life, it has become a means of cultural expression and connection among Newfoundlanders. Today, all but one radio station in the St. John's radio market plays Newfoundland music, suggesting that radio is an important medium used to express cultural identity.

In the St. John's radio market there are eleven radio stations. Six are commercial stations, two are Canadian Broadcasting Corporation (CBC) stations, two have licenses owned by churches, and one station is owned and operated by Memorial University of Newfoundland. All stations broadcast Newfoundland content to some extent except one. The exception is CBC Radio Two that broadcasts only national programs and airs no locally produced content, and was therefore not included in my study.

I interviewed twenty-two station personnel across ten radios station using a semi-structured interview format. I prepared questions that I asked each individual, but followed up with other questions based on participants' responses. Most of my questions sought to shed light on the process by which Newfoundland music received airtime on these stations. Responses to one question in particular, though, revealed that radio stations in the St. John's market served as a cultural connection between Newfoundlanders living in mainland Canada and those at home. I asked, "What can you tell me about the listeners to your show/station?" Not only did participants talk about listeners within physical proximity of the stations' terrestrial signals, but it was through this question that I realized the importance of Newfoundland music to those living off the island.

Interviews with station personnel addressed the producer side of the radio industry. In this study, I also wanted to examine Newfoundland radio from the consumers' perspective. A listener was defined as one who makes a conscious decision to listen to the radio at least two times a month.[10] I used snowball sampling[11] to select twenty-seven people who became my participants.

I used a phenomenological interview process as a research methodology with these listeners. Phenomenological interviews are designed to explore a participant's experience with the phenomenon under examination. Furthermore, time and space are considered in phenomenological interviews because participants are treated holistically. In other words, phenomenological interviews consider the participant's

---

10   With the exception of Radio Newfoundland, which plays 85 percent Newfoundland music, and the stations that occasionally throw in a song by a Newfoundlander during prime time broadcasting, most of the Newfoundland music (by station definition) on the radio is played during a particular show at a specific time. This implies that most of those who listen to the show do so for a specific purpose and not just simply for background music.

11   Snowball sampling, or "identifying cases of interest from people who know other people with relevant cases" (Hay 2000, 44), is acceptable in phenomenological research because the researcher must identity individuals who not only have experienced the phenomenon under study, but are also willing to talk about their experiences with it (Creswell 1998; Thomas and Pollio 2002).

comments within the context of the environment (Peacher 1995). I was interested in listeners' experience listening to Newfoundland music on the radio. In this way, my data reflect participants' perspectives rather than my own.

Listener interviews were also subjected to group analysis.[12] Scholars from the Phenomenology Research Group at the University of Tennessee read through transcripts out loud together, and we constructed a diagram that showed the relationship among emergent themes. In this way, multiple perspectives were used to interpret the data. Five themes emerged from this analysis. The theme of connection between the island and the mainland through radio listening is discussed in this chapter.

Internet broadcasting in the St. John's radio market

In St. John's, Internet radio serves as a means of connection in two ways: first, between the station personnel and listeners living in diaspora, that is between the producer and consumer sides of radio, and secondly, among listeners themselves.

Themes of connection between station personnel and listeners living away

As a result of Newfoundland's history of economic fluctuation, part of the Newfoundland experience for many is having to leave the island, and the radio stations accommodate this trend by broadcasting over the Internet. All of the radio stations in the St. John's radio market offer simultaneous web broadcasts. While only one listener I interviewed listened to radio on the Internet, station personnel told me that the Internet radio broadcasts were highly important to the Newfoundland diaspora. The listeners I interviewed were all physically located within the range of the terrestrial signals of St. John's stations; however, station personnel told me that usually the Newfoundlanders living "away" provide a considerable amount of feedback for their programs by email. Ken Ash, Program Director for Radio Newfoundland, told me from where some of the listeners to the almost-all-Newfoundland music station come.

> Most of our web listeners are not in Newfoundland, as you would imagine. We
> get a lot of Newfoundlanders who work in Ontario and Alberta. There seems
> to be some group of them in Houston as well that you always hear about. I

---

12   Before I conducted phenomenological interviews, a colleague at my institution conducted a bracketing interview with me. The purpose of this interview was to uncover personal biases that might influence my interpretation of the data. The interview transcript was subjected to analysis by a group of phenomenologists who revealed that I had some clear assumptions regarding the importance of my projects, and that I had to be careful not to unintentionally lead listeners to talk about culture and identity, but rather let these themes emerge on their own.

think they're in the oil industry. But, you know, we get them from everywhere ... Alberta, obviously. Ontario ... there's probably more Newfoundlanders in Ontario than anywhere, but where ever work takes them, like Texas. A lot of Newfoundlanders are working in Korea and other places in the Far East as English teachers. And we get some now because there is a Newfoundland post-secondary institution located in Doha, Qatar. There are a lot of Newfoundlanders over there that are teaching and they listen ... We've had some from New York as well ... New York, New Jersey ... could come from anywhere (Ash 2005).

Most listeners tuning in from away feel a stronger connection to home when they hear Newfoundland music outside Newfoundland. One radio station program director allowed me to read some of the emails she received from listeners living away. For example, a Newfoundlander in Ottawa wrote, "My computer at my office and home are now locked in right here [to the Newfoundland station]. With the city going a hundred miles an hour around me, your site takes me back to a different time and place." Another listener wrote in after a friend sent him the link to a particular radio station's listen-live site: "I am so happy. Now I can hear some good tunes instead of this city crap where I live ... I got [the link] yesterday at noon and still never turned it off. It was on all night." A third Newfoundlander proclaimed, "You will never know what its like to be connected to my culture like this. We appreciate it here in Ontario." In other words, radio via the Internet serves as a cultural connection for those living off the island.

This feeling of connectedness goes both ways. Not only do listeners living away feel connected to Newfoundland through Internet broadcasts, the Internet also connects local listeners to their non-local family. "In Radio Newfoundland's case," Brenda Silk (2005), a DJ at Radio Newfoundland told me, "it's a manner of tying [local listeners] together with their loved ones that are far away." Brian O'Connell (2005), Program Director at OZ-FM adds: "It's an opportunity for you for a few minutes to escape that place and come back here [to Newfoundland], at least in spirit." During my fieldwork, Alberta was mentioned as a popular place from which radio stations received calls and emails.

The Internet serves several purposes for radio stations. Besides providing a means for broadcasting outside the range of terrestrial signals, the Internet is also how some stations send and receive digital files of music that they put on the air. DJ, Brad McDonald, explained this procedure.

We work off a server. It's called DMDS. It's digitally mastered music that is posted on a site, which you log into. You've got a user name and a password, and record companies [post music] ... The quality is impeccable. And then I take it from [the site] and just dump it onto our software that we're using (McDonald 2005).

Not only commercial radio uses the Internet to transfer or obtain digital music files. Christine Davies, music librarian for CBC radio in St. John's, told me how beneficial digital music has been for the CBC. The music library in St. John's is a regional library, "a regional source [of music] for the entire province and part of the [CBC] network as well ... If there's a programmer in, say, Alberta, who requires a piece of music, and we're the only library that has it, I arrange a feed to that library" (Davies 2005). The CBC music library in St. John's connects Newfoundland to the rest of the country through its role in the CBC network.

Instant messenger is another Internet service that is useful to radio DJs in Newfoundland. Josh Jamieson and Katie Norman, hosts of *Fresh Focus*, a show about issues concerning Newfoundland's youth, use the Internet to communicate with listeners during their show. "We've set up an MSN Fresh Focus account so we can talk to the listeners online. We've had listeners from all over the world, which is pretty cool" (Jamieson 2005).

Themes of connection and community between listeners

The previous section described the local-global connection perceived by station personnel in the St. John's market. Music and radio also connect Newfoundlanders living on the island to members of the Newfoundland diaspora.

Radio has long played a significant role connecting Newfoundlanders with the rest of the world. The most common historical example of this role mentioned by listeners was the influence of the American radio stations that were present in Newfoundland as a result of the American military bases on the island. World War II, in particular, was a time when radio was important for communication not only on the war front, but on the home front as well. "I think the importance of the radio connecting people to the outside world was tied into that World War II experience," one listener told me. "Many Newfoundlanders that fought overseas were from rural communities, and the only source of information about what was going on was from the radio" (Keough [no relation to the author] 2006). Another listener talked about listening to the radio during the Vietnam conflict.

> I can remember at some point back probably in the early seventies, when the Vietnamese conflict was on, just being so worried ... the fact that when something would happen we'd get the report on it so quickly on the radio and be able to hear it. I remember being so conscious of the importance of it being able to disseminate information, uh, I guess such a great distance, such a short period of time (K. Coffey 2006)

Still another listener recalled how important radio was to her grandfather for hearing information from the rest of the world.

> My grandfather [was] in the house when the news would come on, [and] we all knew, as young children, that we had to be really quiet ... because the radio was the focal point at that time. My grandfather was listening to his news ... 'cause it was very important to him to find out exactly what was on the go in the world,

and we weren't to talk or make a lot of noise at that time or prevent him from listening to that (C. Coffey 2006).

Radio, in these cases, was a way of obtaining information.

American radio stations in Newfoundland during World War II were also glimpses into American culture for many Newfoundlanders. "I used to listen to the American station from the base, too, on Saturday mornings," another listener shared. "I didn't listen to them for music, [but] they'd have cowboys, the Lone Ranger, and all these" (C. Moran 2006). Here, the listener had access to radio programs produced in the United States, so the radio served as a connection to American culture.

American music, however, did not leave Newfoundland when the war ended. American produced music has always been available through the radio. "There was a lot of American music," one listener recalled.

> A lot of it was pop … a lot of it was what I call 'mainstream country,' which, in many cases, was sort of Top-40, but it was Marty Robbins. It was Johnny Cash. It was people like that. We didn't hear a lot of traditional American music (Purcell 2006).

Today, information from the rest of the world can be accessed more easily, and the sense of isolation in Newfoundland is not as strong. Nevertheless, music and radio still play a prominent connective role. This connection has become increasingly important as many young Newfoundlanders leave the island for mainland Canada in search of employment. One listener offered an explanation for why leaving the island is a core theme in Newfoundland music.

> When you long for something, maybe that opens up a floodgate of creativity or something … it has to. Because when you're in a situation and you're living in an environment, you take it for granted. Don't you? It's only when you go away from it that it becomes … something that you almost have to have, like breath itself. You have to have it. I hear it in my daughter's voice. She's out in Calgary. "Mom, I can't wait to get home. Mom, oh my God, I'm so excited! I can't wait to come home" (Mooney 2006).

Another listener in her twenties talked about going to Germany with her fiancée, who is also a Newfoundlander, and missing home.

> When we went away to Germany, we craved Newfoundland music. When you go away, you really crave it, I think. We had one CD with us, which was the Irish Descendents, and we played that over and over and over, and it really makes you homesick … When people say, "Oh, I don't like Newfoundland music," I'm like, "Just go away from home and then you'll enjoy it. You'll appreciate it more when you're not there" (Brenson 2006).

While this listener does not specifically mention radio as a means of connection to her culture, she emphasizes the importance of Newfoundland music in that connection. The Irish Descendents, a popular Newfoundland band, receive significant air time on Newfoundland radio stations, implying that other listeners feel a cultural connection when listening to this group.

While leaving the island is part of the Newfoundland experience, so is returning.[13] This act of leaving and returning has impacted the music because musicians who leave the island return with new influences that often changes Newfoundland music. "These are guys who went to work in Toronto, started to sing, and made records ... and they became quite popular across Canada. Harry Hibbs did and Dick Nolan, you know" (White 2006). In other words, the musicians who left Newfoundland and recorded or performed in other parts have Canada have helped create national and international recognition for Newfoundland as a place that turns out good music.

During the last decade, St. John's radio stations' websites have become portals for connection between listeners in diaspora and those in Newfoundland. One listener told me that the radio program *Jiggs and Reels* on OZ-FM, which is broadcast through both terrestrial and Internet radio, was a point of connection between herself and her fiancée who was doing research in Europe.

> When Stephen was away in Germany ... I used to be on MSN [Instant Messenger], and I used to email "Jiggs Breakfast" [another name by which the *Jiggs and Reels* radio program is known]. And he'd [her fiancée] would turn it on, and it meant something. Sunday morning when Newfoundland music was on, and he was away, I used to try to make a point to remind him that it was on. It was the middle of the day in Germany, so he didn't have to get up early to listen to it, or anything like that. [One time] I was on MSN and he was on MSN, and I emailed Tony Hann and I said, "Can you play 'Salt Water Joys'?" and I gave him my last name and I told him it was for my fiancée who's in Germany doing his research. And Stephen [had also] emailed Tony Hann and asked him to play the song for me. Anyway Tony had a great chat about that on the radio (Brenson 2006).

Here, the radio, Newfoundland music, and the live DJ all help to connect two people living an ocean apart. The locally based DJ, music, and girlfriend interact with the distantly positioned boyfriend through the Internet radio medium.

---

13   In addition to migration, other themes in Newfoundland music include emotional connections to the ocean, methods of surviving harsh winters, relationships to animals, drinking and time with friends, and fishing.

Discussion and conclusions

When I was in Ottawa in the winter of 2006, I had a chance to interview Pierre Louis Smith, then the Vice President for Radio at the Canadian Association of Broadcasters. "Radio is local media," he told me. Pierre was not surprised to hear about local content on Newfoundland radio stations. "Radio cannot compete with the delivery of music over the Internet, so they have to go to their strength, which is local programming and local content" (Smith 2006). This statement from the public agency that represents, among others, all the radio broadcasters in my study tells a lot about the nature of music and the media today. Technology can help with local content, as seen by the testimonies about how important it is for the Newfoundland diaspora to be able to access Newfoundland radio stations. However, technology also can hurt efforts to emphasize local content. The Internet, for example, presents a significant challenge to radio because, at least in Canada, the Internet is not regulated. Listeners can use the Internet to bypass the radio to get the music they want. The key to maintaining a healthy balance between terrestrial radio and its competition is for terrestrial radio to remain flexible. Radio stations in St. John's, have been able to maintain this necessary level of flexibility. In Newfoundland, much of the music by Newfoundlanders broadcast on the radio is about Newfoundland as a place, and about common experiences that Newfoundlanders share. In this way, Newfoundland music has become part of the cultural identity of Newfoundlanders. Therefore, not only is the survival of terrestrial radio in Newfoundland aided by the efforts of station personnel to broadcast local content, the local content *in the music* aids in the survival of these radio stations as well.

Internet radio broadcasts are not the only means Newfoundlanders living in diaspora have of connecting with home. Connections with friends and family on the island are maintained through phone conversations, E-mail, remittances, visits, newspapers, and (to a lesser extent) television.[14] Radio, however, has an advantage over these means of connection: it is a medium conducive to the transmission of music; and music, with its strong Newfoundland themes, is an important expression of Newfoundland identity.

Geographer Kevin Robins criticizes new media technologies, including radio, for shifting their focus away from local or national issues towards global issues (Robins 1995). In Newfoundland, however, I see a trend towards the incorporation or fusion of global issues with local content because migrating (and returning) is both a local and a global phenomenon. The CBC is a good example of this fusion. While the CBC Radio One station in St. John's airs national programs, there is significant airtime dedicated to locally produced programs and programs that focus

---

14 Newfoundland does not have a substantial television production industry. The local CBC television station and NTV (Newfoundland Television) provide news coverage, but few if any other local programs receive television airtime, and these channels are difficult to access outside Newfoundland.

on local issues. Three of the CBC employees I interviewed hosted such programs. Specifically, Angela Antle's *Weekend Arts Magazine* merged Newfoundland content with content from the rest of the world. As the show's host, she added her own personal, local perspective to the broadcast. The CBC model allows for the co-existence of local, national and global material. Listeners who mentioned that they tune in to CBC programs listened because they like to learn what is going on around Newfoundland and also in the rest of the world. For them, the CBC station broadcasts become that bridge.

Radio is a mass medium that crosses both regional and national boundaries (Abu-Lughod 2003). While the Canadian government is concerned about the domination of its media by the United States, the transnational nature of media means that people living in the United States can access Canadian radio as well. Richard Sims, Director of the Montana Historical Society, wrote about how listening to Canadian radio stations became a part of his drive along Montana Highway 2, a stretch of road approximately 50 miles from the Canadian border. He felt that the long drive was best accompanied by the CBC station out of Regina, Saskatchewan. "Hearing Satchmo's jazz set introduced by a French Canadian DJ helps the driver focus on the endless two-lane horizon," he wrote (Sims 2006).

This transnational nature of radio is further evidenced by new technologies to disseminate radio broadcasts. Satellite radio plays a large role in the transnationalization of radio, as American subscribers to XM or Sirius satellite radio receive 10-15 Canadian produced channels of the approximately 120 channels offered. At the same time, Canadian subscribers have access to all the American-produced channels in addition to the Canadian ones, thanks to the agreement between these companies and the CRTC. Listeners can also download broadcasts of radio programs onto their MP3 players, devices which are then transported by their owners to many places, both locally and internationally. In other words, satellite radio, the Internet, and podcasting have becomes conduits through which the international crossover and exchange of cultural material occurs, thus allowing these technologies to become global in their influence.

The influence of new broadcasting technology helps ensure the survival of terrestrial stations in Newfoundland. Web broadcasting and podcasting have allowed those living outside Newfoundland to keep in touch with home. Thus, technology cuts both ways: local content can trump placeless substitutes *and* it can become globalized to the Newfoundland diaspora. Globalization and technological innovation have resulted in the expansion of a community or a nation's geographic limits (Morley and Robins 1995). This expansion is certainly true for Newfoundland. The diffusion of Newfoundland culture was originally limited to the physical movement of Newfoundland migrants. Today, radio bridges physical distances to bring aspects of Newfoundland culture to consumers all over the world. While still quite small, there is now a world market for Newfoundland culture, and radio helps to meet that demand. Through web broadcasts of local material, the terrestrial stations of the St. John's radio market have become transnational entities.

References

Abu-Lughod, L. (2003), "Asserting the Local as National in the Face of the Global: The Ambivalence of Authenticity in Egyptian Soap Opera," in A. Mirsepassi et al. (eds.).

Anderson, B. (1983), *Imagined Communities* (London and New York: Verso).

Ash, K. (2005), Program Director, Radio Newfoundland, Interview.

Berland, J. (1994), "Radio Space and Industrial Time: The Case of Music Formats," in B. Diamond and R. Witmer (eds.).

Brenson, S. (2006), Radio listener, St. John's, NL, Interview.

Canadian Content and Culture Working Group (1995), *Canadian Content and the Information Highway* <http://www.pch.gc.ca/pc-ch/pubs/ihac/3_e.cfm>, accessed 6 June 2008.

Castells, M. (2000), *The Information Age: Economy, Society and Culture, Volume I: The Rise of the Network Society* (Malden, MA: Blackwell Publishing).

Coffey, K. (2006), Radio listener, St. John's, NL, Interview.

Creswell, J. (1998), *Qualitative Inquiry and Research Design: Choosing Among Five Traditions* (Thousand Oaks, CA: Sage Publications).

Crisell, A. (1994), *Understanding Radio*, 2nd Edition (London: Routledge).

Davies, C. (2005), Music Librarian, CBC Radio, St. John's, Interview.

Diamond, B. and Witmer, R. (eds.) (1994), *Canadian Music: Issues of Hegemony and Identity* (Toronto: Canadian Scholar's Press).

Dicken, P. (2003), *Global Shift: Reshaping the Global Economic Map in the 21st Century*, 4th Edition (New York: Guilford Publications).

Grossberg, L., Wartella, E. and Whitney, D. (1998), *Media Making: Mass Media in a Popular Culture* (Thousand Oaks, CA: Sage Publications).

Hay, I. (ed.) (2000), *Qualitative Research Methods in Human Geography* (South Melbourne, Australia: Oxford University Press).

Hay, J., Grossberg, L. and Wartella, A. (eds.) (1996), *The Audience and its Landscape*, 1-5 (Boulder, CO: Westview Press).

Hendy, D. (2000), *Radio in the Global Age* (Cambridge, UK: Polity Press).

Hiller, H. and Franz, T. (2004), "New Ties, Old Ties and Lost Ties: The Use of the Internet in Diaspora," *New Media and Society* 6:6, 731-52.

Information Highway Advisory Council. (1997), *Preparing Canada for a Digital World: Final Report of the Information Highway Advisory Council* (Ottawa: Industry Canada).

Jamieson, J. (2005), DJ, *Fresh Focus*, Radio Newfoundland, Interview.

Johnston, R., Taylor, P. and Watts, M. (eds.) (1995), *Geographies of Global Change: Remapping the World in the Late Twentieth Century* (Oxford, UK: Blackwell).

Keough, B. (2006), Radio listener, Ferryland, NL, Interview.

Keough, S. (2007), "Constructing a Canadian National Identity: Conceptual Explorations and Examples in Newfoundland Music," *Material Culture* 39:2, 43-52.

Lysloff, R.T.A. and Gay, L.C. Jr. (eds.) (2003), *Music and Technoculture* (Middletown, CT: Wesleyan University Press).

McDonald, B. (2005), DJ, Hits 99.1, Interview.

McGee, T. (1985), *The Music of Canada* (New York and London: W.W. Norton and Co.).

Mirsepassi, A., Basu, A. and Weaver, F. (eds.) (2003), *Localizing Knowledge in a Globalizing World: Recasting the Area Studies Debate* (Syracuse, NY: Syracuse University Press).

Mooney, A. (2006), Radio listener, Ferryland, NL, Interview.

Moores, S. (1993), *Interpreting Audiences: The Ethnography of Media Consumption* (London: Sage Publications).

Moran, C. and K. (2006), Radio listeners, Pouch Cove, NL, Interview.

Morley, D. and Robins, K. (1995), *Spaces of Identity: Global Media, Electronic Landscapes and Cultural Boundaries* (London: Routledge).

Norman, K. (2005), DJ, *Fresh Focus*, Radio Newfoundland, Interview.

O'Connell, B. (2005), Program Director, OZ-FM, Interview.

Peacher, R. (1995), *The Experience of Place* (Knoxville: University of Tennessee, Ph.D. Dissertation).

Purcell, B. (2006), Radio listener, St. John's, NL, Interview.

Robins, K. (1995), "New Spaces of Global Media," in R. Johnston, P. Taylor, P, and M. Watts, (eds.).

Royal Commission of Government (2003), *Renewing and Strengthening Our Place in Canada* (St. John's, NL).

Siklos, R. (2007), "Is Radio Still Radio If There's Video?," *New York Times* 14 February 2007.

Silk, B. (2005), DJ, Radio Newfoundland, Interview.

Sims, R. (2006), "Canadian Border and Montana," *Helena Independent Record* 31 December 2006.

Smith, P-L. (2006), Vice President for Radio, Canadian Association of Broadcasters, Interview.

Stanbury, W. (1998), "Canadian Content Regulations: The Intrusive State at Work," *FraserForum* August 1998. (Vancouver, BC: The Fraser Institute).

Statistics Canada (2007a), "Labour Force Characteristics by Sex and by Province, 2006," *Canada Yearbook*. (Ottawa, ON: Minister of Industry).

—— (2007b), "Population by Province and Territory 1861-2006," *Canada Yearbook* (Ottawa, ON: Minister of Industry).

—— (2007c), "Alberta's Population Boom," *Canada Yearbook* (Ottawa, ON: Minister of Industry).

Stern, J. (2003), *The Audible Past: Cultural Origins of Sound Reproduction* (Durham, NC and London: Duke University Press).

Thomas, S. and Pollio, H. (2002), *Listening to Patients: A Phenomenological Approach to Nursing Research and Practice* (New York: Springer Publishing Company).

Turk, E. and Johnston, D. (1997), *Competitiveness, Access and Canadian Content: The Three Pillars of Canadian Internet Policy* (Ottawa: JFK School of Infrastructure Project).

Webb, J.A. (1994), *The Invention of Radio Broadcasting in Newfoundland and the Maritime Provinces 1922-1939* (Fredericton and Saint John: University of New Brunswick, Ph.D. Dissertation).

Wong, D. (2003), "Plugged in at Home: Vietnamese American Technoculture in Orange County," In R.T.A. Lysloff and L.C. Gay, Jr. (eds.).

US Bureau of the Census. (1997), Current Population Reports 1997: Computer Use in the United States, <http://www.census.gov/prod/99pubs/p20-522.pdf>, accessed 6 June 2008.

White, J. (2006), Radio listener, Ferryland, NL, Interview.

Chapter 12
Local Independent Music Scenes and the Implications of the Internet

Holly C. Kruse

Almost two decades ago, I began research on local independent ("indie") rock and pop music scenes. Specifically, I examined issues of identity, social and economic networks, social interaction, place and space, and gender in these cultural, social, and economic formations. I did ethnographic and archival research, and I focused on two local indie scenes: one in Champaign-Urbana, Illinois—a college town which at that time had a vibrant indie pop scene—and San Francisco (Kruse 1993 and 2003). Since I did my research, the Internet has become a key player in the production, promotion, dissemination and consumption of independent music. Before discussing the ways in which the Internet has changed the landscape of indie pop and rock music, focusing on some of the key findings of my earlier Champaign-Urbana case study may help illustrate some general trends in indie pop and rock music scenes from the mid-1980s until the mid-1990s.

The Champaign-Urbana indie scene in the early 1990s

Champaign-Urbana, Illinois is home to a large state university and was an active site of indie pop and rock music production during the 1980s and 1990s. Champaign is a fairly typical college town, and unlike Athens, Georgia or Austin, Texas, Champaign-Urbana had not occupied a central position in the history of music-making practices and/or styles outside the mainstream music industry. Despite this, Champaign-Urbana spawned many indie pop/rock bands that were regionally popular, and in a few cases, nationally known, including, in the 1980s, the Elvis Brothers, Turning Curious, the Farmboys, Weird Summer and Combo Audio. By the early 1990s Champaign was home to several alternative acts that had signed major label deals—such as Poster Children, Adam Schmitt, and Titanic Love Affair—and to small independent labels like Parasol and 12 Inch Records.

Parasol was part of a growth in independent labels in the United States that began in the 1980s and that was accompanied by an increase in the number of bands recording their music (as opposed to simply performing it live). Many of these artists recorded for very small labels that were unable to pay them advances before recording; however, larger indies like Touch and Go and Frontier were able to fund the production of records and videos. Very few indie pop or rock bands in communities like Champaign-Urbana recorded for large indies, though; many more released records for smaller indies like Parasol.

Parasol existed in the 1990s at the intersection of two distinct vectors: 1) the emergence of the 7-inch single as an important marketing tool in indie music, especially, indie pop; and 2) a specific history of indie music production in Champaign-Urbana. The rise of the underground 7-inch market coincided with the mainstream music industry's determination that the vinyl phonograph record was no longer a profitable, and therefore viable, format. Independent labels, and especially smaller indies, were reluctant to abandon the cheap vinyl format for CDs, since many did not have the available capital to invest in CD production and most indies did not want to be relegated to the production and distribution of cassette tapes. Undoubtedly, part of vinyl's appeal to independent companies selling alternative rock and pop was that the major labels' effort to make the format obsolete, in effect made whatever appeared on 7-inch vinyl to be in opposition to the mainstream. Furthermore, the format catered to a rather select audience: those who still owned, or were willing to purchase, turntable technology.

Prior to the advent of Parasol in its early indie-single-selling incarnation, and throughout the 1980s, a number of indie pop and rock bands appeared in Champaign-Urbana, and several released albums, singles, and/or tapes on local labels like Office, Trashcan, and Popsicle. The existence of small local labels, and the availability of relatively cheap analog recording equipment (usually four- or eight-track), enabled bands to make recordings available locally without relying on signing major label or major indie deals. According to Trashcan's founder, its entire purpose was "to break the local scene" (quoted in Springer 1989, 8).

Indeed, a 1989 local newspaper article declared "Champaign-Urbana is on the verge of becoming a trend-setting music scene with national influence" (Springer 1982, 8). While this optimistic prediction did not exactly come to pass, several local artists were signed to major label or major indie deals, including Poster Children, which released records on indies Frontier, TwinTone, and Sub Pop, and then moved to Sire; Hum, which signed to RCA, but released records on independent labels like Dedicated, 12 Inch Records, and Mud (a Parasol-affiliated label); and former Champaign musicians Ric Menck and Paul Chastain, whose band Velvet Crush recorded for Warner Brothers in the United States and Creation in Britain.

The sense that Champaign-Urbana was a regional scene of some national significance was undoubtedly important to the visibility of labels like Parasol, yet Parasol had not had as much success selling records in Champaign-Urbana as it did through mail order. For instance, singles by Champaign music scene veteran Nick Rudd did not sell well in Champaign: his first single on Parasol sold eighty copies in England and one at the record store at which he worked. While those in the greater indie pop subculture saw Parasol as important in getting local music to the public, Parasol founder Geoff Merritt argued that despite a roster laden with local talent, he was not doing them much good in the local area: "I'm just putting out singles, and nobody's buying them" (Merritt 1991). He recalled a time in the early 1980s when there was more of a sense of community in the Champaign scene:

> It used to be a single came out in this town and everybody bought it. "Stabs in the Dark" [a 1982 compilation album of Champaign indie pop and rock bands] came out and everybody bought it whether they liked the stuff or not, because it was local ... I guess there's something wrong with putting out 45s because a lot of people don't even own turntables anymore, but even so, people should buy this stuff (Merritt 1991).

Thus, it was at a moment in the early 1990s that the Champaign scene was most in the national spotlight that one of its key participants saw the scene as least cohesive.

For instance, one near non-participant in the local indie pop/rock scene had historically been the University of Illinois's student-run radio station, WPGU. From its establishment in 1967 and lasting into the 1990s, WPGU was an album rock station, and this made the station a site of contention. Indeed, a local musician remarked in the early 1990s, "I'm always so blown away when I go to another town and I hear their college radio station, because PGU is—I hate to get on anyone's case—but they're really awful" (Schmitt 1992). In the early 1990s, WPGU switched to a "top of the alternative charts" format.

However, most music scene participants saw this shift to another commercial format as not particularly adventurous, and thus, for most Champaign-Urbana listeners who wanted to hear non-formatted alternative music radio shows, the only option was the community radio station, WEFT. As a community radio station, WEFT's overall philosophy was (and is) to provide the community with programming that was not otherwise available, and in Champaign-Urbana, this included independent pop and rock. Within its programming mix in the late 1980s and early 1990s, alternative rock and pop occupied about 20 percent of WEFT's slots. WEFT also devoted slots to a number of other types of music and programming that were not available on radio stations in the listening area, including alternative news and information, world music, bluegrass, folk, and jazz. For listeners looking for non-major label alternative rock/pop on the radio in Champaign-Urbana, WEFT could be a rather frustrating source of material. As a local musician noted, WEFT was "so sporadic, it's hard to know when you turn it on what you're going to be hearing" (Schmitt 1992).

Live performance venues had also been a source of frustration for musicians and other music scene participants. By the early 1990s there were two clubs in Champaign-Urbana that booked indie acts, Mabel's and the Blind Pig, but most local indie musicians complained that Mabel's, the larger of the two venues and the one that was located near campus, was essentially closed to them. This had not always been true; during much of the 1980s Mabel's was the primary performing venue for local alternative acts. However, in the late 1980s and early 1990s the club had booked more mainstream local acts, prompting a record label owner to state that "Mabel's doesn't book bands like ours for various reasons" (Merritt 1991). However, a local musician added that with the opening in 1990 of the Blind Pig, a club located away from campus, "it's been both easier and better for bands to find a place to play" (Schmitt, 1992). Still, the Blind Pig was often criticized for its small size, heat, and location; and the limited number of venues meant that Champaign-Urbana musicians often found their hometown to be one of the harder markets to enter.

The importance of local identity in indie music in the 1980s and 1990s

Indie music, more than most forms of pop and rock music in the United States, had in the 1980s and 1990s been identified by locality (Athens, Seattle, Austin, Minneapolis, Champaign, Olympia, and so on), both by participants and by those outside particular scenes; therefore, the way in which indie music was and is understood in relation to notions of local identity merits examination. Identities were (and are) formed, changed, and maintained within localities that were constituted by geographical boundaries, by networks of social relationships, and by a sense of local history.

Placing one's participation within this context is a way of asserting the importance of one's position in local music history. For instance, a particular Champaign musician constructed his involvement in local music as pioneering by articulating his band's relationship in time and space to other bands and local scenes; he claimed "I think we were one of the first completely original local alternative college bands, after the Vertebrats—there was probably no one before them. We were sort of paralleling what the Replacements were doing in Minneapolis without even knowing or hearing of them" (Gerard 1991). By locating his band on a level of importance similar to that of a seminal local band and a nationally prominent band, the musician identified his band as one of local and possibly even national importance, at least within the confines of his narrative.

Indeed, music scene participants in many places in the United States were, in the early 1990s, for the most part aware of some version of local music history and placed themselves within that tradition, whether it was in Champaign's indie pop scene, San Francisco's punk scene, or Seattle's grunge scene. Participants were part of social formations in which existing musical practices and traditions affected emerging music.

Yet identification of a locally-defined "Champaign scene" in the 1990s came at a time when participants were in fact also being connected in some way with trends and entities that transcended locality: for example, Parasol and the other "local" indie pop 7-inch labels, like Washington, DC-based label Slumberland; or Poster Children and other harder rocking indie bands, like Minneapolis's Soul Asylum. Touring played an important role in creating cross-scene relationships among participants. On the level of scenes, social interaction transcended geography. A Champaign indie musician explained:

People come to me asking about certain people in certain towns. Tonight I'm going to go see Die Kreuzen, who we always used to stay with every time we played Milwaukee. If they were in town, they expected to stay with us. We're part of a group where you see someone every six months or every three months (Gerard 1991).

Interlocal networks such as these, because they brought institutions and people in disparate local scenes together in broader systems of cultural production and dissemination, underscore the degree to which economic structures of indie music were, in the 1980s and 1990s, interrelated in numerous ways with social practices.

Indie music's local scenes and socio-economic structures in the last part of the twentieth century could, in the end, be seen as overlapping networks, in which musical knowledge, genre, geography, and position in the independent music business located subjects within one or more networks. Shared musical knowledge and practices were important in the formation and maintenance of interlocal social and economic networks, and thus made it impossible to ever understand a formation like the Champaign-Urbana music scene in isolation. Today, in the age of the Internet, looking at scenes in isolation makes even less sense, as the ability to connect with others across scenes and to disseminate independent music has become easier than ever before.

Locality and interlocality in the Internet age

Certainly, things have changed in local indie music scenes like Champaign-Urbana's with the widespread popularity of the Internet and other home digital technologies. For example, if there is no local broadcast radio station that plays fairly obscure independent music, one can tune into an Internet radio station that does. One can watch video of live shows on YouTube rather than enduring the hassles associated with seeing live music in person. And the Internet, with its ease of connecting people across localities, regions, countries, and continents, may well play a role in a decline in a sense of local identity and of being part of a particular local history within a music scene, and in the growth in the sense of translocal identity. As Internet options for the discussion and sharing of indie music increase, the local spaces devoted to interaction around music may well suffer: one London-based founder of indie labels and of a venue in the San Francisco area observes that one of his favorite local record stores

closed down earlier this year and is now only an online mail order website ... I seem to have lost touch with them since they became online only—I used to love going in to the guys working there and would always end up buying more based on their personal recommendations (Sideboard 2007).

Also in demise is the perception that there are "local sounds" and identities associated with specific locations. In their book *Soundtracks: Popular Music, Identity and Place,* John Connell and Chris Gibson (2002) claim that the spread of the Internet has increased the flow of subcultural music and information across disparate localities, thus helping to "de-link the notion of scene from locality" and that for those involved in little-known music genres, the Internet enables a sense of offline "imagined community" that is crucial to scenes but not tied to geography (107). Even as some local spaces survive, the Internet has likely accelerated the process of regional, national, and international sounds and practices interacting with local music. Increasingly, it is argued, geography doesn't matter. Thanks to the Internet's ability to facilitate "virtual scenes," Andy Bennett and Richard Peterson (2004) state that although virtual scene participants are geographically distant, these individuals "around the world come together in a single scene-making conversation via the Internet" (10).

Still, Connell and Gibson (2002) argue that even in the Internet era of music production and distribution, local spaces and identities remain important, providing the necessary infrastructure still required for music scenes to survive. They add that for many of the musicians they examined, local space affected the music recorded and evoked a sense of place. Indeed, Connell and Gibson report that staying local and playing small club and pub circuits is prized by some bands and labels who believe that these activities connote a degree of "authenticity" that breaking through to larger, more heterogeneous audiences would not. Even if we put authenticity issues aside, physical sites of local music remain important for scene participants. A European indie music fan, and Internet and society scholar, comments that "indie rock concerts are still important, record stores are as well, as places to talk about and listen to music, network with people" (Fuchs 2007). The founder of a small independent label in the San Francisco Bay area also reports little decline in the number of local spaces, like records stores and other informal public places, in which music and knowledge of local music history are disseminated (Mallon 2007). Regarding record stores in particular, the European fan and commentator argues that "indie rock fans don't stop buying music in local indie record stores, they now consume more music, from more different sources" (Fuchs 2007).

The decentralization and globalization of music production, dissemination and consumption that the Internet is credited with fostering has not, apparently, resulted in the disappearance of local identity, local scene history, and even the sense that local sounds exist. A music writer in the eastern United States observes "I still see much evidence of regional pride, not just here, but in other scenes and larger cities as well. It still means something to be a Philly band, or a D.C. band, or a Cincy band, or a Portland, OR band" (Grover 2007). To the extent that a perception persists that the notion of "local sound" is becoming passé because of the effects of the Internet, one San Francisco area musician remarks "It seems to me that the 'death of a local scene/sound' started before the Internet became a force in music though" (Skaught 2007), and another adds "I think the Bay Area has been 'music scene' challenged for about 25 years" (Ray 2007).

Whatever the state of local scenes, it is certainly true that the Internet has played an extraordinarily important role in developing the sort of inter- or trans-local connections that were evident in my research done in the late 1980s and early 1990s, before the Internet was widely used by the American public. As Connell and Gibson (2002) point out, it is now easy for "parallel sub-cultures to become connected" through the Internet and related resources (107). An indie label owner I interviewed agreed that the Internet allows "a disparate fan base to get in touch with each other" (Mallon 2007). In his examination of goth scenes in Britain, James Hodkinson (2004) finds that participants in goth subcultures are able to refer to web pages that serve as clearinghouses for information across localities about events, bands, and places. Because goth is largely a style-based subculture, it is not surprising that Hodkinson finds a particular translocal Internet effect to be the transmission and discussion of information on style and shared subcultural values. At the most basic level, Internet discussion boards and other resources facilitate personal connections across localities. One scene participant in Britain, when asked if she felt less localized because of connections made through the Internet, replies, "Yes, you know people from London, you know people who go to Slimelight [London goth club], you know people from Edinburgh, you know people from Glasgow, you've talked to them" (Hodkinson 2004, 143).

Unlike Hodkinson in his analysis of goth scenes and the Internet, Bennett and Peterson differentiate between online—or "virtual"—scenes and offline "conventional" music formations. They note that while conventional local music scenes involve various live offline events like concerts, "virtual" scenes are comprised of mediated one-to-one communication, largely between fans, which, they argue, makes the virtual scene much more one of the fans' making. As Hodkinson's research demonstrates, and as I will argue later, the conventional and the virtual are not truly separate. It is true, and obvious, however, that music scene participants are now more easily able to access each other and connect with other participants, both nearby and faraway, because of the Internet. Moreover, the Internet helps fans to be in contact not just with each other, but also with musicians. As Marjorie Kibby (2000) observes, having "an electronic place in which to 'gather' enables a direct link between fans, and even makes possible a direct connection between fans and performers" (91). Again such connections are not new, but because of the relatively easy accessibility of individuals through the Internet, digital communication technologies further contribute to the long-accepted common sense in indie music that there are few differences or barriers between musicians and fans.

Local scenes, commerce, and the Internet

Digital communication technologies may also be, as many observers argue, allowing indie musicians and record companies to better reach potential listeners

212 Sound, Society and the Geography of Popular Music

with their music. The Internet helps bands go on tour; one record label owner observes that the bands he knows that tour, "find it easier to find gigs, places to stay, and people to whom to sell tickets and merchandise" (Mallon 2007). Furthermore, record labels that were not previously able to get their product in many record stores, or to many buyers through mail order, or played on the radio, can easily have a presence on the Internet, no matter how small or obscure the label (Connell and Gibson 2002). Nor do they need to sell their music online to financially benefit. Rich Egan, the co-founder of the independent label Vagrant Records claims that "Our music, by and large, when kids listen to it, they share it with their friends ... Then they go buy the record; they take ownership of it" (quoted in McLeod 2005, 529). Unlike the major labels, argues Kembrew McLeod (2005), small independent record companies do not see peer-to-peer file sharing of music as cutting into their business; in fact, they find it a good way to promote their often obscure music genres and bands. But another independent label co-founder and co-owner disagrees about the efficacy of file sharing and mp3 files in general as a way to sell CDs. He states, "One of our artists, Jill Tracy, does very well selling MP3s of individual songs. For the rest, I think MP3s function as try-before-you-buy items and ways for bloggers to spread the word about our CDs. I don't think it sells us many CDs, though" (Mallon 2007).

Whether file-sharing directly results in CD sales or not, the technology has been integral in disseminating more local music to regional, national, and international audiences than ever before. In addition to the Internet, a newer technology that has been important in getting local music to larger publics is digital recording technology. Musicians can now, relatively inexpensively, make professional quality recordings even if they don't have the financial backing of a record company to pay for studio time or are located far away from a recording studio (Connell and Gibson 2002). As discussed earlier in this chapter, reasonably inexpensive home recording devices are not new: in fact, they played a key role in the growth of indie pop and rock music during the 1980s and the embrace of "low-fi" recording quality as a marker of "authenticity." But digital technology creates a cleaner, more professional sound, and Connell and Gibson maintain that the further spread of home recording facilitated by digital technology "in many metropolitan areas has suggested the potential for decentralization, through cheaper and more accessible technology, Internet resources, and capabilities for global distribution and marketing for unsigned bands" (2002, 258). Indeed, McLeod adds that "Today, there is the very real possibility that most musicians can make a living from a small but loyal fan base, and completely bypass the bloated entertainment industry" (2005, 530). A Milwaukee music writer I interviewed lists the various ways that the Internet and related technologies have allowed musicians to reach fans with high quality recordings and circumvent the conventional music recording and delivery systems:

> In the past, gatekeepers controlled access to recording equipment, to recordings themselves, and to distribution and promotion. The internet, in some ways,

obviates all of that (certainly the first two). It's theoretically possible now for a band to form entirely online (whether locally or not), record its songs entirely on computer-based software, and (this is probably the key step) through buzz built by big mp3 blogs (notably Fluxblog), end up with a recording contract and a fair amount of fans (Norman 2007).

A world in which any local musician can reach any fan anywhere on the globe with his or her music would indeed lead to a radical decentralization of music production and distribution, likely combined with a decreasing emphasis on local knowledge, history, connections, and material resources. A musician could live in the mountains of Tibet and have access to the same information technology and channels as one located in Los Angeles or London, and his/her location would be, in theory, of little or now importance.

Such thinking, however, is indicative of a belief in what media scholar and political economist Vincent Mosco (2004) calls "the digital sublime": the complex of myths surrounding the Internet. As Mosco puts it, the Internet provides us with

> ... a story about how ever smaller, faster, cheaper, and better computer and communication technologies help to realize, with little effort, those seemingly impossible dreams of democracy and community with practically no pressure on the natural environment ... Moreover, the story continues, computer networks offer relatively inexpensive access, making possible a primary feature of democracy, that the tools necessary to empowerment are equally available to all. Furthermore, this vision of the Internet fosters community because it enables people to communicate with one another in any part of the world (2004, 30-31).

In fact, we know that access is not available to all, that digital technologies of recording and distribution may be relatively inexpensive for some but still very expensive for many others, and that gatekeepers still exist in the Internet era.

It is true that traditional media and structures of the mainstream music industry are now less important in determining the music that reaches the public than they were prior to the late 1990s; however, there are now new gatekeepers. In a universe in which a computer user can potentially access thousands and thousands of songs, how do people know where to find music they like? New digital recording and distribution technologies mean that local musicians are competing with thousands of other DIY local musicians to sell their records online, or to be noticed on MySpace, or to have their mp3s downloaded, or to have their songs offered on eMusic (or, if signed to a cooperating major record label, iTunes). Connell and Gibson point out that with the advent of music distribution on the Internet "Unless musicians [with web pages] could generate significant links from other websites, or could mobilize audiences for self-promoted materials, their sounds were likely to be lost in a 'sea' of digital noise" (2002, 261). Indeed, despite the great optimism about the ability of the Internet to circumvent gatekeeping apparatuses of the mainstream industry, many music listeners may be turning to other gatekeepers, like iTunes's, Amazon's, and other online commercial behemoths' links to what people who bought a particular song or CD also bought, or to what an algorithm has determined that the consumer might like. Indie artists in local scenes still struggle to be heard, despite the decentralizing technological forces that have emerged in the past 10 to 15 years.

Of course, the problem for musicians and small labels of having one's music heard via the Internet, and for listeners of finding music that they like on the Internet, are problems specific to those who have affordable, high-speed Internet access. It is wonderful to have one's music played on Internet radio, and to hear new music on Internet radio—to give one example—but what about those in the United States who cannot afford a broadband connection, or an Internet connection of any type, or who lack access to a computer? For them, Internet radio is irrelevant. And what about musicians and potential listeners outside of the developed world, the vast majority of whom cannot afford computer access, and even if they could, may well lack the infrastructure—electricity, an Internet connection—to go online?

The decentralization of music potentially allowed by the Internet requires widespread network connections, and the time and resources to use the connection for disseminating and/or downloading or streaming music. The result is, according to Connell and Gibson, "a selective geography at the global scale" (2002, 263). As an international scene participant I interviewed put it, "I see that while the Internet should afford local scenes greater attention, that instead it seems to reinforce the big global hubs" (Borschke 2007). Thus, the Internet, and the transference of aspects of music scenes to it, combined with barriers to access and the resulting national and international digital divides, make problematic assertions about the Internet's ability to provide local music, including indie music, with vast new and translocal audiences. There is no doubt that the Internet has increased music's ability to transcend geography, but as more independent record labels move exclusively to the Internet-even if they are still primarily selling CDs, which do not require a computer to play, and not mp3s—recreational high-speed Internet access may become crucial in participating in music scenes. That kind of access was, as of 2007, present in less than half of American homes (Horrigan and Smith 2007).

Offline interaction and locality

So, it seems that various forces and relationships are likely in many places to conspire to keep much music and music interaction local. And for those who have Internet access, the Internet not only allows more interaction between and among geographically distant music scene participants, as discussed earlier in this chapter, it helps create face-to-face interaction between and among geographically proximal participants, in more or less organized ways. Music writers may find themselves targeted by local bands seeking publicity. One states "even though I have never limited myself by any means to writing about local bands, these are the bands most likely to seek me out to send promos, invitations to shows, and MySpace friend requests" (Grover 2007). Obviously, however, the Internet has proven a useful tool for increasing face-to-face opportunities for all kinds of music scene participants, not just writers. A Boston-area musician and writer says, "Craigslist [a service that provides free online classified advertisements and discussion forums] is an amazing tool for hooking musicians up. I joined the Hyphens as a consequence of searching for people looking for bass players in the Boston area" (Mayo-Wells 2007). He adds

> Also, in two cities now – DC and Boston – I've observed/participated in online forums centered around local print publications that became nuclei for (aspects of) the local music scene – providing a moderately incestuous pool for band-member swaps, opportunities for inter-band networking and cross promotion. On both of these forums, you'd often see messages of the "so, which shows is everybody going to tonight" form (Mayo-Wells 2007).

This phenomenon is observed on a slightly larger geographic level by Hodkinson of British goths, of whose online conversations he writes "rather than removing the need for physical travel, the tendency was for such virtual interactions to encourage goths to want to see their friends in face-to-face circumstances" (2004, 143).

That the Internet may largely be useful for creating and maintaining contacts in music scenes that also involve face-to-face connections should not be surprising to those familiar with the research on local Internet networks. In their study of an entirely wired housing development in a Toronto suburb—a development that they call "Netville" in which every home has free, high-speed Internet access—Keith Hampton and Barry Wellman (2002) find that the great majority of Internet users use the Internet to communicate with those they already know, and that users who do form online relationships often take these relationships offline. The latter observation is especially relevant to relationships in local music scenes, which may be formed more easily among those who first "meet" online than in the physical spaces of scenes, like clubs and record stores, where various barriers to meeting and getting to know people—the awkwardness of approaching strangers in public, the uncertainty that people in these places share one's interests, the possibly diminishing number of certain kinds of sites like record stores, the background noise of clubs—make undertaking the task of creating personal connections prohibitive.

Of further interest, Hampton and Wellman found in Netville that "wired" residents knew the names of 25 neighbors, while those who did not go online and use community Internet resources knew the names of only eight neighbors. Also, "wired" Netville residents had more face-to-face contact in their neighborhoods: they regularly talked offline to twice as many neighbors as their "wireless" counterparts and visited each others' homes 50 percent more often that those who were "wireless." Clearly, the Internet can lead to a greater network of weak-tie local connections for those who use it, and its local resources. Even absent explicitly social-networking sites like

MySpace and Facebook, the Internet can and does bring otherwise unconnected community members into contact—including those involved in local music—and increase one's sense of neighborhood and local identity.

Conclusions

The widespread dissemination of music and technological processes decentralizing music production have been underway for centuries. The exposure of local and regional music to geographically distant audiences has not always served to lessen the importance of locality and local identity-in fact, sometimes such dissemination has the opposite effect. The popularity of the phonograph and recorded music in the early twentieth century meant that, for instance, blues music from different cities could be recorded, shipped, and sold all over the country. When one could buy records of New Orleans blues music in St. Louis, and vice versa, a real sense of what was New Orleans blues versus St. Louis blues versus Chicago blues developed. Broader dissemination helped to create and reinforce ideas that there were distinct local sounds and local identities.

New communication technologies therefore do not necessarily cause the death of the local. Further careful analysis needs to be done to understand the complex interaction among local indie music institutions, individuals, histories, and sites and emergent technologies that allow for some, but hardly all, greater access and connection to people and resources across local music scenes. Given the history of utopian narratives about new communication technologies creating global communication and understanding, and the continuing existence of local and regional cultures and of barriers to technology and information access, the complete erasure of physical space in subcultural music identities, histories, and institutions is not likely to happen anytime soon.

References

Bennett, A. and Peterson, R.A. (eds.) (2004), *Music Scenes: Local, Translocal, and Virtual* (Nashville: Vanderbilt University Press).

Borschke, M. (2007), An International Perspective on Local Scenes [email correspondence] to Kruse, H.C. [9 May 2007].

Connell, J. and Gibson, C. (2002), *Sound Tracks: Popular Music, Identity and Place* (London: Routledge).

Fuchs, C. (2007), Independent Music Scenes and the Internet in Europe [email correspondence] to Kruse, H.C. [17 June 2007].

Gerard, D. (1991), Experiences in the Champaign-Urbana Music Scene [interview by H.C. Kruse] Champaign, Illinois, 19 December, 1991.

Grover, J. (2007), Music Writing and Independent Music Scenes [email correspondence] to Kruse, H.C. [11 May 2007].

Hampton, K.N. and Wellman, B. (2002), "The Not So Global Village of Netville," in Haythornthwaite and Wellman (eds.).

Haythornthwaite, C. and Wellman, B. (eds.) (2002), *The Internet in Everyday Life* (Malden, MA: Blackwell).

Hodkinson, P. (2004), "Translocal Connections in the Goth Scene," in Bennett and Peterson (eds.).

Horrigan, J.B. and Smith, A. (2007), "Home Broadband Adoption 2007," *Pew Internet & American Life Project*, <http://pewinternet.org/pdfs/PIP_Broadband%202007.pdf>, accessed July 2007.

Kibby, M.D. (2000), "Home on the Page: A Virtual Place of Music Community," *Popular Music* 19:1, 91-100.

Kruse, H. (1993), "Subcultural Identity in Alternative Music Culture," *Popular Music* 12:1, 33-41.

—— (2003), *Site and Sound: Understanding Independent Music Scenes* (New York: Peter Lang).

McLeod, K. (2005), "MP3s are Killing Home Taping: The Rise of Internet Distribution and its Challenge to the Major Label Music Monopoly," *Popular Music and Society* 28:4, 521-31.

Mallon, J. (2007), 125 Records [email correspondence] to Kruse, H.C. [19 June 2007].

Mayo-Wells, D. (2007), Experiences in East Coast Music Scenes [email correspondence] to Kruse, H.C. [9 and 11 May 2007].

Merritt, G. (1991), Parasol Records and the Champaign-Urbana Music Scene [interview by H.C. Kruse] Champaign, Illinois, 19 December, 1991.

Mosco, V. (2004), *The Digital Sublime* (Cambridge, MA: The MIT Press).

Norman, J. (2007), Music Writing and Local Scenes in the Midwest [email correspondence] to Kruse, H.C. [11 May 2007].

Ray, G. (2007), A Musician's Experience in Northern California Music Scenes [email correspondence] to Kruse, H.C. [8 May 2007].

Schmitt, A. (1992), Experiences in the Champaign-Urbana Music Scene [interview by H.C. Kruse] Champaign, Illinois, 17 June, 1992.

Skaught, B. (2007), Experiences in the San Francisco Music Scene [email correspondence] to Kruse, H.C. [8 May 2007].

Sideboard, T. (2007), Independent Pop Music in the U.K. [email correspondence] to Kruse, H.C. [18 June 2007].

Springer, P.G. (1989, February 17), "Back to the Garage," *Champaign-Urbana News-Gazette Weekend*, 8-9.

Chapter 13
Where Are the New US Music Scenes?

Ola Johansson and Thomas L. Bell

Introduction

In 1981, a relatively unknown band called REM issued their first record and within a few years had achieved the status as the most innovative and successful rock band in United States. As a result of REM's success, Athens, Georgia became the archetype of the college music scene (Jipson 1994). Ten years later, in the Pacific Northwest, Nirvana played their guitars with a raw intensity not heard since the glory days of the punk revolution. To Nirvana it smelled like teen spirit, to the music industry it smelled like money, and to the media it smelled like grunge. Predictably, the criticality of Seattle, grunge and Kurt Cobain eventually faded (Bell 1998). Using a popular culture yardstick, 1991 is a long time ago, which brings us to the question: Where are the *new* music scenes in the United States?[1] Our sense is, at least judging from reading the music press during the last few years, that no scenes as recognizable as Athens and Seattle have emerged. Of course, all cities have music scenes to a greater and lesser extent; however, we will take a broad approach to investigate this question. The basic method employed in this research is to identify the place of origin of a sample of new artists that have developed a certain level of prominence and popularity nationwide. The type of music we investigate is alternately called "modern rock," "alternative rock," or "college rock." These categories have arisen mainly for the purpose of radio station formatting, and the terms are used more or less interchangeably as catch-all phrases for multiple subgenres of new rock music that exists in opposition to bestselling Top 40 pop music. Henceforth, we will, for the sake of simplicity, refer to this broad category of popular music as "new rock." The method of tying artists to places of origin is, admittedly, a rudimentary way of identifying scenes, and some of those limitations will be discussed further in the research methods section of this chapter, but our data also include information on music genres and record labels to broaden the discussion of scenes. Before the patterns of new rock music are presented and analyzed, however, the concept of a scene must be discussed further.

---

1   The data that we analyze is not limited to artists from United States; many originate in Europe and Canada, for example, but the lion's share of the chapter is dedicated to identification of scenes and spatial variability of new rock in the United States.

The nature of music scenes

Broadly speaking, a scene involves the production and consumption of specific musical styles with collectively shared tastes and activities (Cohen 1999). A scene can be internally tight-knit with very strong personal and musical connections among the artists involved, or it can be only loosely connected. The *ideal* of a scene is represented by the tight-knit scenario, but in reality, most scenes, when scrutinized more closely, do not exhibit, at least in an artistic sense, such cohesiveness. Moreover, scenes can be separate, overlapping or nested within each other. For example, many music genres are artistically related to each other and distinct boundaries may be hard to identify, thus creating overlapping scenes. In other cases, genres may be categorized as sub-genres of broader scenes within which they are nested. From a temporal standpoint, scenes are also unstable as they constantly appear, disappear, and change (Harris 2000). Finally, and most importantly for geographic analysis, scenes range from local, i.e. based on proximity, to non-local, in which fans and performers of a genre are scattered geographically but cohesive in their allegiance to the genre through the Internet or in disparate live performance settings.

If we assume that most new rock music scenes, at least partially, are local, it can be useful to frame the existence of scenes around the notion of place-based creativity. By place-based creativity we mean that some places, because of local cultural and socio-economic circumstances, are more likely than others to be centers of new rock music. To use a specific, localized example, in Richard Florida's book *The Rise of the Creative Class* (2002), the author encounters a heavily pierced, "punky" looking top graduate on his campus—Carnegie Mellon University in Pittsburgh—who is recruited by national high-tech firms. The young man eventually settles on an offer from an Austin, Texas-based firm because, in his mind, Austin is *the* cool place to be due to its trendy youth culture, including a thriving local music scene. This vignette does not tell us whether the migrant formed a band of his own (probably not as he was headed towards a career in computer engineering which is, by all accounts, more lucrative and financially secure than the music industry). It tells us, however, something about the processes that geographically concentrate creativity in endeavors that include, but are not limited to, music scenes. This is an uneven spatial process and only selected places give birth to musical innovation—Austin, but less so Pittsburgh, apparently possesses this musical innovation—either fleetingly or by sustaining creative activities for longer periods of time. This place-based creativity is, of course, nothing new. As Peter Hall (1998) points out in *Cities in Civilization*, innovation, whether in the cultural or economic enterprises, has emerged in particular locations throughout history, capable of capturing the spirit of the times. The idea of place-based creativity deemphasizes the internal artistic coherence implicit in the scene concept, and is therefore an attractive point of departure for this research where we employ a macro perspective that cannot explore in detail the connections within individual scenes.

What does it take for a local scene to take shape? To generalize about the formation of scenes, a wide range of factors play a role, but we suggest that all factors can be sorted under one or more of three dimensions: geography, culture, and economy. These interrelationships are represented visually in Figure 13.1 where sixteen factors are positioned in a triangle diagram[2] and they "load" to different degree on the three dimensions. Note that this is a conceptual model only, that the factors are not completely independent from each other, and that more factors can most certainly be added to the diagram. The factors we have identified and shown the numbers in Figure 13.1 are:

1. Impulses from the local *physical environment*, such as the rainy weather in Seattle or the stark landscape of Iceland, act as sources of inspiration for artists. Such explanations are commonplace in media, but rarely given much credence by geographers.

2. *Relative location*: Specific geographic situations are associated with musical innovation. Port cities can be cultural melting pots (e.g. New Orleans), sites for early diffusion (e.g. rock'n'roll from Liverpool), or provide opportunities for musical consumption in live venues. Other cities can be gateways for migration and cultural diffusion (e.g. Memphis's role in the development of blues). Even in times of global informational flows, similar locational arguments cannot be entirely discarded. A Chapel Hill-based band pointed out to us recently that the town's location was ideal as a base for any touring band—good accessibility to both southeastern and northeastern performance opportunities (Two Dollar Pistols 2004). In a related fashion, cities like Salt Lake City and Denver are intermediary places on the circuit pathways for touring bands traveling from the West Coast to the East Coast, or vice versa (Johansson and Bell 2008).

3. *Size of place*: Small places may be more conducive to cooperation and performance opportunities. Athens, Georgia has reportedly benefited from such an environment (Jepson 1994). On the other hand, large cities benefit from localization economies. But, even in larger cities more localized proximity can be important. Scenes can, for example, grow out of "urban villages" centered around a particular venue. The famous club CBGB's association with punk and new wave in New York is an example.

4. The local *socio-economic environment* creates the underlying conditions from which cultural expressions form. For example, the politicized segment of the punk rock movement was intimately tied to deindustrializing places in the UK in the late 1970s.

---

2   A triangle diagram is a graph that can visualize three axes in a two dimensional way. The most common triangle diagram familiar to geographers is probably the soil classification triangle based on three textural classes (sand, clay and loam). The way to interpret the diagram is to follow the dashed lines to each side of the triangle and read how a factor ranks high or low on the geography, economy, and culture dimensions. For example, a factor that is located at the absolute top of the triangle is a purely spatial one.

5. *Strategies of urban development* may influence scenes. In Manchester, the city offered financial incentives, and softened the regulatory demands on the licensing of club venues (Brown, O'Connor, and Cohen 2000; Haslam 2000). Local authorities in Europe tend to have a more active agenda to sustain cultural industries, but also in the United States, cultural openness, diversity, and acceptance of new ideas translate into places where creative people choose to live. Therefore, it is a local policy imperative to nurture such trends. In reality, however, top down control of cultural developments is hard, and gentrifying urban spaces are sometimes forcing out clubs (and thus potential scenes) via regulations or rising real estate prices (Hae 2005).

6. *Music industry infrastructure* includes elements such as record labels, studios, management, and so on. If successful, one particular label or studio can be strongly associated with a city and carry its reputation as a scene. Saddle Creek Records in Omaha, Nebraska has, for example, become a nexus of activity for musical acts that are important to the emo subgenre just as Sub Pop Records in Seattle spawned the grunge movement in the early 1990s.

7. What type of local live scene exists? A multitude of performance *venues* is often crucial. A variation on the venue theme is an annual festival. The *South by Southwest* music festival in Austin, Texas is a key element of that city's reputation as having a strong live music scene.

8. A supportive *local media*, especially alternative press and radio, can be active disseminators of information of a scene.

9. A *culturally open and permissive atmosphere* allows for creative thinking.

10. *Egalitarian social structures* are commonly associated with innovative thinking in the economic realm (Hall 1998) and may be associated with cultural innovation as well.

11. A scene is not only comprised of music production, but also music consumption. A local *audience* exists in a dialectic relationship with local bands. A sophisticated audience creates fertile grounds for new musical expressions as well as impulses from the outside world via artists that are likely to schedule concerts there.

12. In *networks of learning*, local bands know each other, influence each other, even exchange members at times. This can foster musical coherence of a scene, but more frequently, perhaps, such interaction promotes other forms of learning, such as how to successfully maneuver the vagaries of the music industry, promote their music, or master the recording of music.

13. *Individual innovators*: In the inception phase of a scene, both musical innovation and supporting infrastructure often develops from creative individuals and self-made entrepreneurs who subsequently can carry the momentum of a scene forward.

14. Scenes can be influenced by the spatial flows of *tourism*. To satisfy the demands and expectations of visitors, local musical cultures are shaped to offer a locally enhanced experience. This happens in disparate locations and audiences, from global backpackers desire for a world music experience (Gibson and Connell

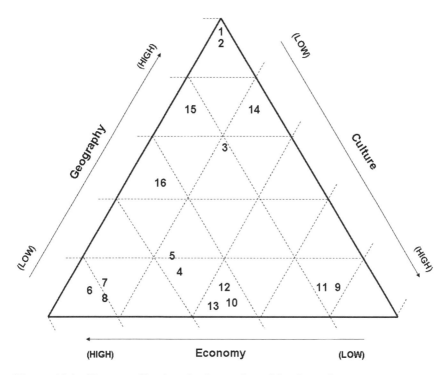

Figure 13.1   Factors affecting the formation of local music scenes

2003; see also Chapter 5 in this volume) to blues and country music in Memphis and Nashville entertainment districts (Johansson 2004).

15. The ethnic and cultural characteristics of immigrants can drastically alter a local music scene, creating new or hybrid musical expressions. College towns experience continuous *migration,* which can sustain a scene, especially when aspiring musicians or fledgling bands relocate to places with a recognized scene.

16. To what extent are places real hubs of musical creativity, or artifacts of the *non-local media*'s need to mythologize local scenes (Mitchell 1996)? Objective musical landscapes exist alongside popular imagined ones that reflect an inherent need to attach spatial meaning to music. This is generally done by the music press, but also through other media, such as film (Wyndham and Read 2003). Media can over-stress the coherence and interconnections among bands emerging from the same place and time, that they are musically interrelated, and that such connections stems from a creative process in which proximity plays a major role. It is also in the interest of the music industry's marketing departments to "invent" scenes. Occasionally, mythologizing about place can come from within. The best example is perhaps traditional New Orleans music and its self-referential nature that reinforces the notion of New Orleans as a "world to itself."

The factors presented above primarily concern the internal dynamics of local scenes. Today, however, the cultural and economic forces of globalization have reworked the relationship between music and place. With postmodern non-hierarchical aesthetics emerging, the geographic analysis of popular music has connected with the "new" cultural geography (Leyshon, Matless, and Revill 1998; Connell and Gibson 2003). One set of ideas refers to Theodore Adorno's (1976) notion that popular music is inevitably moving towards a state of mass production, standardization, and homogenization. But from the vantage point of the early twenty-first century, the movement towards conformity only tells half the story of contemporary popular culture that is also characterized by heterogeneity and a multiplicity of subcultures. Today, this resulting subcultural fragmentation is sometimes referred to as "neo-tribalism" in which individuals search for community and belonging among people with whom they identify (Maffesoli 1996). The division of popular music into distinct subcultures is such an important cultural identifier that musical taste becomes a central element in the formation of identities that signify wider worldviews and political attitudes.

This fragmentation contradicts the idea that globalization also means homogenization because such subcultures, which sometimes find their expression in local music scenes, are continuously and dialectically coexisting alongside more mainstream expressions of popular culture. If rock music was truly homogenized, it would result in a product devoid of place-based content and meaning. But in reality, fragmentation and subcultures often lead to geographically distinct forms of music.

From an economic perspective, the production of music as a post-Fordist cultural industry has a propensity to thrive on geographic agglomeration as well as informational and economic connections in increasingly global networks (Scott 2001). The former indicates that there is a local component to music production while the latter suggests a global tendency. Both are true; five major labels dominate the industry worldwide, while production remains a locally contingent activity. In the global-local nexus the forces of capitalism work themselves out in different ways in different places. This is conceptualized as fluidity and fixity by Connell and Gibson (2003). "Fluidity" indicates the geographic flows of music and the mobility of individuals engaging in musical expressions, while "fixity" emphasizes the place-based, either as unique hearths of music or more recent processes where hybrid musical forms become fixed geographically and associated with particular places. Kong (1996, 1997) uses the term "transculturation" to describe the process of cultural interchange where musical performers and styles resist global homogenization by reworking elements of global music into musical forms that contain distinctly local elements. While such tendencies might be most obvious in non-Western cultures, there is no clear connection between innovation in new rock and core-periphery relations *within* Western countries (where virtually all the music under scrutiny in this chapter originates). Hall (1998) argues that innovation might, in fact, benefit from a geographical location that is peripheral with respect to sites of power and previously dominant economic and cultural paradigms. In modern rock, the affluent periphery of Iceland, Australia, and Sweden has recently been very productive (e.g. Power 2003; Connell and Gibson 2003). The same appears to be true within the United States. Localities where scenes emerge do not necessarily "confluence with the hierarchy of capital or investment opportunity" (Bell 1998, 41). The industry's A&R people often seek out talent in its own geographic territory, chasing creativity wherever that might lead them, which means that musical creativity is spatially separate from the sites of power in the music industry.

Considering the impact of globalization and communication technologies on music scenes, three general outcomes are possible: local, translocal (a network of local scenes) or virtual (which only exists in forums such as fanzines or Internet networks) scenes (Peterson and Bennett 2004). One could argue that in a globalized popular culture where creative input is as likely to come from worldwide networks as from the immediate surroundings, there may be less room for distinctly local scenes. Recent studies have shown that emerging translocalism characterize, for example, the Goth scene (Hodkinson 2004) and the extreme metal scene (Harris 2000). The Riot Grrrl scene, while it emerged in Olympia, Washington and Washington DC (Leonard 1998) as a movement, developed as a feminist ideology without collective spaces, thus exhibiting tendencies ranging from the local to the virtual.

Despite these new developments, "the particularities and histories of each [local] scene" (Kruse 1993, 39) can explain why a specific genre finds fertile ground locally. The identity and taste of individuals are formed locally, even if similar tendencies can be found in many different places. Cultural and socio-economic forces themselves are translocal, and the receptiveness toward specific musical trends can also exist at multiple locations, even if they may be traced to a particular place of origin. Straw (1991) believes that college scenes are the same everywhere; the same culture replicated across the country without much differentiation. Such a position, in our view, over-generalizes the level of homogenization created by national and global forces. Cultural and socio-economic forces are not entirely universal, and spatial variation in the adoption of specific scenes, or new rock in general, is therefore to be expected. Locally contingent forces articulate global trends into viable and distinct local scenes that subsequently influence musicians and audiences in other places. In order to investigate empirically if such local differences exist in the United States, we now turn to the data for further analysis.

Research method

Although theoretical advances have been made, along with case studies of particular scenes, no systematic research has attempted to broadly assess the origin of new rock. In this chapter, we analyze artists using the ten most played college radio albums from the *College Music Journal (CMJ)* as the data source.

Table 13.1   Genres of new rock

Genre	Description	Typical artists from data
Adult alternative	Adult alternative is positioned slightly outside mainstream popular music and generally appeals to an older demographic compared to other genres discussed here. The category is musically eclectic.	Ani DiFranco, Tom Waits, Tori Amos
Americana	Americana is roots-based music adapted to contemporary sensibilities with influences from country, blues, bluegrass, rockabilly, and folk.	Old 97s, Calexico, Southern Culture on the Skids
Electronica	Electronica refers to electronic music that ranges from techno (see Chapter 15 in this volume) and rave music designed for the dance floor to more ambient music with less emphasis on straightforward beats and rhythm.	Moby, Morcheeba, Tricky, Cornershop
Emo	Emo originated as a branch of hardcore punk, but as of late most bands that are labeled emo have adopted a more melodic mainstream style. Emos are mopey punk rockers that dare to talk about (or at least sing about) their feelings.	Get Up Kids, Jimmy Eat World, Braid
Garage	Garage rock is a raw and simplistic form of rock, often with bluesy undertones. It originated during the 1960s with bands that made records on the cheap. The style has gone through several revival phases, including the 2000s when the genre achieved a moderate level of commercial success.	Yeah, Yeah, Yeahs, The White Stripes, The Raveonettes
Hard rock/metal	Loud and obnoxious rock music without redeeming social qualities, which is why the fans love it.	Primus, Deftones, Queens of the Stoneage
Indie rock	Indie rock is, in a narrow etymological sense, music on independent record labels. However, we use the term more loosely as "the core of alternative rock," including bands that are not clearly associated with other specific sub-genres.	Spoon, Constantines, Modest Mouse

Genre	Description	Examples
Indie pop	Similar to indie rock but with stronger pop sensibilities and a guitar-based sound that is sometimes described as "jangly." Indie pop bands have an affinity for sixties pop and they all idolize the Smiths.	Belle and Sebastian, Wolfie, The Charlatans
Post-rock	While embracing the instrumentation of rock, post rock eschews the chorus/verse structure and the "banal" rhythm of traditional popular music, and it is therefore more experimental in character. Although lyrics are sometimes featured, post rock does not shy away from instrumentals.	Tortoise, Sigur Ros, Spiritualized
Power pop	The musical arrangements of power pop are sparse but tasteful with pronounced melodies and vocal harmonies. Power pop also embraces guitar riffs, but without resorting to excessive solos. As such it is inspired by classic pop with a hint of hard rock.	Matthew Sweet, Superdrag, Fountains of Wayne
Psychedelic/retro	Music that more or less faithfully recreates the styles from the late 1960s and early 1970s that drew inspiration from the trippy effects of illegal pharmaceuticals.	Luna, Olivia Tremor Control, Dungen
Punk	Punk rock ranges from nihilistic to anarcho-leftist in character, but it is always anti-establishment. Musically, the short, fast, simple songs originated as an antidote to boring mainstream 1970s music. Moreover, punk values a do-it-yourself ethos over formal musical skills.	Sleater-Kinney, Bad Religion, A Perfect Circle
Rap/hip hop	Hip hop features rapping to electronic music that is often constructed as a collage by DJs. Rap is a form of chant more than actual singing that emphasize rhythmic and often improvised, semi-autobiographical lyrics. See also Chapter 16 in this volume.	Outkast, De Las Soul, Bloodhound Gang

The list emphasizes new musical acts, but also includes some established artists that maintain creative output that continue to have contemporary relevance. *CMJ* was selected because it provides a comprehensive and well-balanced picture of new rock in the United States and also the foreign artists that have an impact on new rock in the United States. Almost all college radio stations report their airplay to *CMJ*, and as these reports form the basis for the chart, no regional bias should exist in the data. Artists that appeared from 2000-2005 are included in the analysis. In all, 294 artists appeared 688 times on the list. Performers that originate outside North America have not been mapped, and have not been accorded the same level of scrutiny in the analysis section. Based on the *CMJ* list, we developed a database that includes the following variables: artist, artist's origin, musical genre, and record label. The chart position was not recorded as the purpose was only to establish a comprehensive sample of new rock artists. The supplemental information was compiled from online music resources, especially biographical sketches and album reviews from rollingstone.com and cmj.com.[3] In some cases, the evolution of a band is a complex process and it is not possible to determine one single "point of origin." Such artists are recorded as having more than one "hometown." The goal is to establish where an artist developed and became part of a scene. Thus, the database is not a list of artists' birthplaces, nor is a later career move to cities such as New York or Los Angeles considered. Moreover, artists are aggregated at the metropolitan level. For example, artists that hail from Hollywood or Anaheim, California are classified as Los Angeles artists. Genre classifications are inherently subjective, often disputed, not to mention resisted, by the artists themselves. Nevertheless, categories are useful guiding instruments, and for the purpose of this chapter, we used genre categories developed at rollingstone.com. The web site provided a very large number of musical categories and in many cases classified individual artists as belonging to more than one category.[4] The elaborate and detailed categories from rollingstone.com are grouped to produce a more manageable number of genres. The genres, genre descriptions, and examples of performers in each genre are shown in Table 13.1. Finally, the headquarters location of record labels (the location of the individual labels themselves, not their potential corporate owner) is mainly determined using the music industry sites www.discogs.com and www.allrecordlabels.com.

The geography of new rock

The data indicate that the emergence of new performers is a spatially uneven process. Figure 13.2 shows the origin of new rock artists by city. In absolute terms, the New York area predictably stands out with 34 artists in the database.

---

3   We would like to thank Amanda Price at the University of Pittsburgh at Johnstown for compiling a portion of the data.

4   This genre classification feature is no longer available online.

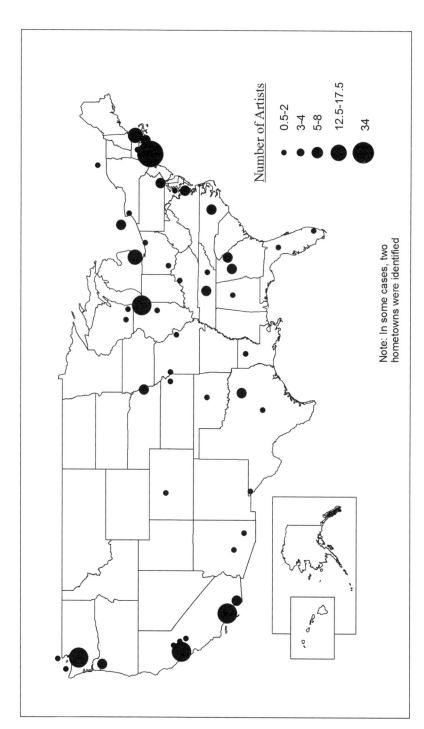

Figure 13.2   Origin of new rock artists by city © Glynn Collis

A cluster of cities distantly follows New York: San Francisco (17.5 artists), Seattle (15 artists), Los Angeles (14 artists), and Chicago (12.5 artists). Forty-eight additional cities are represented by eight or fewer artists. When adjusted for population size, New York is still over-represented as a source of new rock, but the two cities with most per capita artists are Seattle and San Francisco. Less prolific, but nevertheless important sites of new music—Portland, Oregon (4 artists) and San Diego (4 artists)—make the West Coast the artistic epicenter of new rock in the twenty-first century. On the East Coast, the northern tier of Megalopolis—Boston (8 artists) and Providence (4 artists) in addition to New York—perform a similar function, while further south, Philadelphia, Washington DC, and Baltimore together only produced six artists. In the South, college towns are featured on the map, especially Chapel Hill, North Carolina and Athens, Georgia, which, for a long time, have been associated with alternative music scenes. In Athens, a network of musicians referred to as Elephant 6 that both recorded together and formed other bands carries the town's reputation as a center for independent music.[5] Although Athens is the center of the Elephant 6 collective, members are active around the country. Some larger southern cities are also well represented, most notably Atlanta, Dallas, and Nashville. Standouts further north include Detroit and Toronto. Canadian artists mainly appeared towards the end of the research period, a trend that appears to have accelerated, and, if the data extended to the present, it is possible that Canadian cities would have been more prominent.

On the other hand, many sizeable cities have no artist on the list—the largest of those cities is Houston, Texas (Figure 13.3). The geography of the "no shows" is quite distinct. Almost all of Florida's larger cities are in the no artist category. Extending westward across the Gulf Coast, a broader region of low creativity in the new rock field is detected. The industrial heartland stretching from Pittsburgh into the Midwest is a similar void for new rock. If we include Kansas City and St Louis, cities where only a couple of artists hailed from, a Midwestern "trough" is evident. Note that this region does not include the cities of the Upper Midwest/Great Lakes. Finally, a third "underperforming" region is, surprisingly, the Southwest where cities like Las Vegas, Salt Lake City, Denver, and Phoenix are home to few artists. A few comments can be made about these patterns. Neither Florida nor the industrial Midwest are demographically conducive to youth-based cultural creativity. Both regions have older-than-average population structures. Some recognized southern centers of rap music—Miami, Houston, New Orleans—are more or less absent from the new rock map; the two musical styles appear to exist in geographic opposition to each other in the South. Moreover, southern places known as historic centers of innovation in music that is derived from African-American culture—Memphis and New Orleans—are absent from our list.

If no local scenes existed, all cities could be expected to produce the same number of artists per capita. Under such circumstances, a simple correlation analysis of city

---

5   Bands that are associated with Elephant 6 include The Apples in Stereo, The Olivia Tremor Control, Neutral Milk Hotel, Of Montreal, and others.

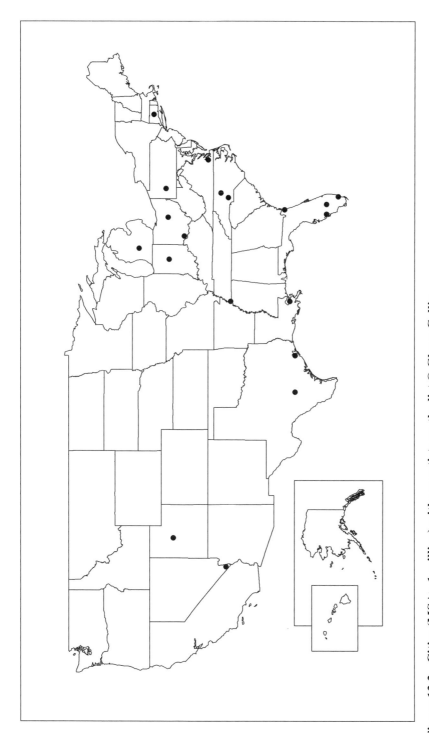

Figure 13.3   Cities (MSA> 1 million) with no artists on the list © Glynn Collis

population size and number of artists would yield an r-value of 1.0, while a lower value indicates some level of geographic concentration. The actual correlation is .83 for the cities that are represented in the database. (The correlation would be weaker if all cities with no artists in the data are included.) The result is close enough to 1.0 to suggest, in addition to the already identified clusters of musical creativity, that a relatively dispersed spatial pattern exists, which could indicate that trans-local and virtual scenes may be important forces at work. To better understand the connections both within local scenes and among different locations, the character of the music must be investigated further. Therefore, we grouped the artists into thirteen genres as explained in the methods section (see Table 13.1).

The Americana category is geographically spread out but does generally not originate in college towns. As a musical category that draws from American folk traditions, especially in the South, it is not a surprise that Americana music exhibits a pattern that is different than many other genres of new rock. However, Americana is not strongly represented on college radio, so the results here must be considered highly tentative. The categories "electronica" and "post-rock" show strong similarities. Both are strongly international in character (in fact, two-thirds of the artists in these categories combined originate outside North America) and although a recent study points to post-rock having strong virtual scene characteristics (Hodgkinson 2004), artists in these categories originate mainly in large cities in the United States, especially New York City and Chicago. The latter is associated with a particular label—Thrill Jockey—that emphasizes these musical genres. This pattern indicates that global musical sub-cultures take hold in large cities that are better connected and more receptive to such trends. This is probably a two-way flow between global cities in the new rock world, such as New York and London. The fact that we see different type of scenes—local, trans-local, and virtual—in these niche genres is not surprising or contradictory. Relatively small music subcultures can only develop local scenes in large places with a critical mass to sustain them, while other places are reliant on communication through virtual networks.

Although slightly less internationalized, rap/hip hop is also found in big cities on the East Coast or the West Coast. This bifurcated pattern reflects a common geographic distinction in rap/hip-hop where East Coast rap draws from its longer history of rap centered in New York and the West Coast is associated with "gangsta rap" (Krims 2000; Keyes 2002). There are only a modest number of rap/hip hop acts that receive air play on college radio, so much like the Americana genre, the pattern here should be interpreted with caution. In fact, the connection between rap and place in the United States has been described more extensively elsewhere (e.g. Forman 2002). Some smaller categories such as "garage" and "psychedelic/retro" are also highly international, although associated with particular world regions, such as Scandinavia and Australia/New Zealand. In North America, these genres can be found in different locations—from big cities to college towns, and from the East Coast to the West Coast. The fact that such music appears in disparate locations suggest that they have tendencies toward translocalism. On the other

Table 13.2   Genres by size of city

Electronica	4.7
Hard Rock/Metal	4.5
Rap/Hip Hop	4.4
Post Rock	4.4
Punk	3.7
Adult Alternative	3.4
Indie Rock	3.1
Americana	2.9
Emo	2.8

Cities are organized in five categories by size of metropolitan region: 5. >8 million, 4. 4-8 million, 3. 2-4 million, 2. 1-2 million, 1. < 1 million. Note: Some genres are excluded due to the small sample size.

hand, the "emo" scene is entirely a United States phenomenon. It has found fertile grounds in most parts of United States (except Los Angeles, according to the data). Emo draws its energy from the disgruntlement of the emo artists and their audience with suburban generic conformity (Bell and Bell 2002). The stifling and decidedly uncreative environments of suburbia created the angst of the affluent white teens that drive this musical movement, and as such suburban environments are ubiquitous and placeless (Relph 1976), emo logically has translocal tendencies, which the data also indicate. The musically related "punk" category (which includes "old-school" punk that is faithful to the original sounds of the 1970s, "pop punk," and "skate punk" rather than the introspective emo version of punk) is, on the other hand, strong in Los Angeles and the West Coast, as is the hard rock/metal scene. The California punk scene is a continuation of the early local punk movement as the music has been institutionalized via labels and networks (Gosling 2004) which enables this musical form to continue to thrive in this location. The largest genre is "indie rock," which is a broad category that contains alternative rock that does not fit into any of the other more specific categories. Subsequently, it exhibits a pattern that mirrors the overall tendency of the data: much from the coasts—San Francisco, Seattle (but no artists from Los Angeles!) and the Northeast. For example, all artists from Boston falls in the indie rock category.

From the data analysis above, it is not only evident that there are regional patterns, but also that the size of cities is a variable that influences the development of new rock. To investigate the relationship between genres and city size, we divided cities into five size categories and then calculated the category from which the typical band in each genre originated (Table 13.2). City categories by size were created because the dramatic variation in absolute size between cities makes means unreliable and could distort the results. Four genres show an association with relatively large cities. Electronia and post-rock were recognized earlier as cutting edge genres that emerge in large cities. The hard rock/metal category is found high in the urban hierarchy because of its long connection with Los Angeles. Rap/hip-hop has through its diffusion pattern a strong connection with the largest cities in the United States. On the other hand, the genres that show a propensity to develop in more moderate-sized places include emo, Americana, and indie rock. An important explanation for the result on indie rock is the genre's connection with college scenes, which often, but not always, are located in relatively small cities.

But for new rock scenes as a whole, how important is the existence a university? Our data indicate that a small minority of artists actually originate from college towns. We have not attempted to define exactly what constitutes a "college town," but even if we use the term liberally and include mid-size cities that may be viewed as having college-oriented music scenes—such as Seattle, Boston, and Minneapolis—only 22 percent of the acts are from college towns. At the same time, a band from a "non-college city," such as New York, may have formed in college or have ties to a local college scene, but our data are not capable of making that distinction. To equate new rock primarily with college scenes is, however, clearly an oversimplification.

Another recognized component of local scenes is music industry infrastructure, especially record labels. The geography of record labels in the database shows far greater concentration than that for artists. The three largest US cities—New York, Los Angeles and Chicago—dominate the labels in descending order. The roster for labels in the three major cities is made up of approximately three-quarters non-local artists (104 out of 140 are from outside the metropolitan area). These labels clearly cast a wide net in their search for talent around the country. New York labels in particular issue releases from international artists and therefore act as "ports of entry" for new musical trends.

All other locations together have a greater association with local talent—almost half of the musical acts (21 out of 46) come from the home region of the label. Some labels pursue a narrow artistic focus. In such cases, they have the capacity to be identified as part of, or even create, a local scene. Examples include the punk label Kill Rock Stars in Olympia, Washington, and Thrill Jockey in Chicago, which is dominated by post-rock artists. Saddle Creek in Omaha, Nebraska emphasizes indie rock, but it mainly draws attention through its most successful artist, Conor Oberst's Bright Eyes. The fact that Bright Eyes call Omaha home is undoubtedly the reason why the city registered at all in our data as a place of origin for new rock. Moreover, some college towns have labels that play an important role in the formation of local scenes, even if they include many non-local artists in their rosters. Examples of such labels are Kindercore in Athens, Georgia, Lookout in Berkeley, California and Merge in Chapel Hill, North Carolina.

Discussion

The data have revealed the spatial characteristics of new rock and its several subgenres. Within the United States, the size of urban centers plays an important role defining the character of local scenes. Most artists come from sizeable cities, the overwhelming majority from metropolitan areas larger than one million people. Most small towns that produce important new rock are college towns, although college towns do not dominate the geography of new rock. Moreover, the regional dimension of new rock scenes is relatively strong. The West Coast and the Northeast are the most prolific regions, while at the other end of the spectrum, some regions appear to lack the cultural dynamic to produce much that is noteworthy, including Florida, parts of the Midwest, and the Southwest. Thus, there are clear geographic differences in place-based creativity within the context of new rock. There may be underlying social and cultural forces that strongly influence the pattern as noted in this research. The most creative cities in the United States, as listed by Florida (2002), correspond closely with the centers of new rock. For example, the two cities we dubbed as the most productive rock cities are number one (San Francisco) and number five (Seattle) on Florida's creative list.[6] Another example of how broad societal differences within the United States must be taken into account when discussing the emergence of new rock is the contemporary political map of the nation. The "blue-red" divide is at least partially replicated in the map of new rock where the West Coast (especially from San Francisco and northward) and the Northeast are liberal-Democratic *and* places where new rock thrive. The Upper Midwest is also more Democratic than the "Lower Midwest" and greater contributor to new music scenes. The South, on the other hand, cannot as easily be explained in this manner. These broad cultural factors are usually not considered when an explanation for the emergence of local scenes is sought (Figure 13.1), but we do believe that regional differences in new rock fit into the patterns of creativity and culture in the United States.

To expand on such geographical differences, not only production patterns, but also consumption patterns should be investigated further. *Rolling Stone* magazine recently published a guide rating the best places to pursue a college education if you are a music fan (Eliscu 2005). Consumption-oriented variables such as venues, record stores and radio stations are investigated and almost all top ten cities are also prominent in our database—again Seattle is number one, followed by cities such as Chapel Hill, Detroit, San Francisco, Athens, Georgia, and Portland, Oregon—thus, production and consumption of new rock appear to be flip sides of the same coin.[7] Based on personal communication with a staff member of a major contemporary rock band, we sense that not only CD and ticket sales, but also the audience response during a concert—the "going nuts" factor—varies geographically. Lower turnout and more muted responses are typical of

6   Florida's top ten list includes the following cities: 1. San Francisco, 2. Austin, 3 (tie). San Diego, 3 (tie). Boston, 5. Seattle, 6. Chapel Hill, 7. Houston, 8. Washington DC, 9. New York, 10 (tie). Dallas, 10 (tie). Minneapolis.

7   The *Rolling Stone* list includes the following cities: 1. Seattle, 2. Chapel Hill/Raleigh/Durham, 3. Detroit, 4. Austin, 5. Nashville, 6. St. Louis, 7. San Francisco, 8. Athens, GA, 9. Minneapolis, 10. Portland, OR.

Texas and the Southwest, while the opposite is true for the Northeast according to our informant (Koch 2006). Again, audience attitudes (or musical consumption preferences) exhibit some overlap with the geography of music production.

Beyond the peaks and troughs of the production of music, enough of a dispersed pattern exists within the United States to also suggest translocal and virtual tendencies at work. Moreover, the most frequently played albums in some genres come from international artists indicating translocal connections on a global scale among places in the United States and Europe. It is interesting, moreover, that Seattle is still dominant in US music space. Our data are not temporally extensive enough for a time series comparison to explore how the geography of new rock differs from the recent past and, for example, if Seattle occupies a different position now than during the 1990s.

Is there a new scene in Seattle? The music is relatively consistent; the overwhelming number of acts are guitar-based rock with more or less strong ties to the alternative/punk movement (but no rap, electronica, or post-rock). Bands include Modest Mouse, Sunny Day Real Estate, Foo Fighters, Death Cab for Cutie, and Black Rebel Motorcycle Club, so in some ways the answer is yes, there is a Seattle scene, although none of the contemporary bands "define" the scene as Nirvana did. Similarly, San Francisco probably suffers, much like Seattle, from a comparison with its own past where old scenes (especially Ashbury-Haight era psychedelia) have been so cemented in popular music history that little discursive room exists for new scenes. Seattle and San Francisco today are, at best, perceived as containing heightened levels of place-based creativity.

Therefore, we conclude that attention must be paid to practices, particularly in media, that frame the emergence and departure of scenes. The discursive element to scenes has been recognized (Cohen 1999; Peterson and Bennett 2004) as a primary tool to connect places and music. Such a connection is generally made by the media and music industry to authenticate music and to situate it within a particular genre. Especially in "alternative" music, validating the authenticity of the music through attachment to place has been especially important. It is done in opposition to more commodified popular music, which implies detachment from place and therefore less "authenticity." As long time readers of *Rolling Stone* magazine (and other music press), we have detected a lack of recognition of the continuous productivity of cities such as Seattle and San Francisco. In fact, the few recent reports on scenes and the connections between place and innovative music have involved artists and cities that are more "newsworthy." For example, the music of the low-fi blues duo the Black Keys is inspired by the post-industrial landscape of the rust belt city of Akron, Ohio, according to one *Rolling Stone* article (Scaggs 2008), and Baltimore was designated "best scene" as a "hotbed" for art rock (*Rolling Stone* 2008). Note that these locales did not figure prominently in our data. While the authentication of music is still important in the twenty-first century, it is common to do so using translocal and virtual discourses as much as, and perhaps even more than, local scenes. This can explain our initial reflections on how local scenes appear less prominent today.

References

Adorno, T. (1976), *Introduction to the Sociology of Music,* transl. (New York: Seabury Press).

Bell, T.L. (1998), "Why Seattle? An Examination of an Alternative Rock Hearth," *Journal of Cultural Geography* 18:1, 35-47.

Bell, T.L. and Bell, L. (2002), "Is There a Geography of Emotion? Lessons Learned from the Emo Music Culture," Paper presented at the AAG meeting, New Orleans, LA, March 2002.

Bilby, K. (1999), "'Roots Explosion:' Indigenization and Cosmopolitanism in Contemporary Surinamese Popular Music," *Ethnomusicology* 43, 256-96.

Brown, A., O'Connor, J., and Cohen, S. (2000), "Local Music Policies Within a Global Music Industry: Cultural Quarters in Manchester and Sheffield," *Geoforum* 31, 437-51.

Cohen, S. (1999), "Scenes," in B. Horner and T. Swiss (eds.).

Connell, J. and Gibson, C. (2003), *Sound Tracks: Popular Music, Identity and Place* (London: Routledge).

Eliscu, J. (2005), *Schools That Rock: The Rolling Stone College Guide* (Wenner Books).

Florida, R. (2002), *The Rise of the Creative Class* (New York: Perseus Books).

Forman, M. (2002), *The 'Hood Comes First: Race Space and Place in Rap and Hip Hop* (Middletown, CT: Wesleyan University Press).

Gibson, C. and Connell, J. (2003), "'Bongofury:' Tourism, Music and Cultural Economy at Byron Bay, Australia," *Tijdschrift voor Economische en Sociale Geografie* 94:2, 164-87.

Gosling, T. (2004), "'Not for Sale': The Underground Network of Anarcho-Punk," in R. Peterson and A. Bennett (eds.).

Hae, L. (2005), "Zoning Out Dance Clubs in New York City: Neoliberal Urban Development and Alternative Urban Cultures," Paper presented at the AAG meeting, Denver, CO, 5-9 April 2005.

Halfacree, K.H. and Kitchin, R.M. (1996), "'Madchester Rave On:' Placing the Fragments of Popular Music," *Area* 28:1, 47-55.

Hall, P. (1998), *Cities in Civilization* (London: Weidenfeld and Nicolson).

Harris, K. (2000), "'Roots'?: The Relationship between the Global and the Local within the Extreme Metal Scene," *Popular Music* 19:1, 13-30.

Haslam, D. (2000), *Manchester, England: The Story of a Pop Cult City* (London: Fourth Estate).

Hodgkinson, J. (2004), "The Fanzine Discourse over Post-rock," in R. Peterson and A. Bennett (eds.).

Hodkinson, P. (2004), "Translocal Connections in the Goth Scene," in R. Peterson and A. Bennett (eds.).

Horner, B. and Swiss, T. (eds.) (1999), *Key Terms in Popular Music and Culture* (Malden, MA: Blackwell).

Jipson, A. (1994), "Why Athens? Investigations into the Site of an American Music Revolution," *Popular Music and Society* 18:3, 19-31.

Johansson, O. (2004), "Entertainment Districts and Downtown Revitalization in Nashville and Memphis," Paper presented at the AAG meeting, Philadelphia, PA, March 2004.

Johansson, O. and Bell, T.L. (2008), "Touring Circuits and the Uneven Geography of Rock Music Performance," Paper presented at the AAG meeting, Boston, MA, April 2008.

Keyes, C. (2002), *Rap Music and Street Consciousness* (Urbana: University of Illinois Press).

Koch, K. (2006), Webmaster and roadie for the band Weezer. Personal communication, 23 February 2006.

Kong, L. (1995), "Popular Music in Geographical Analysis," *Progress in Human Geography* 19:2, 183-98.

—— (1996), "Popular Music in Singapore: Exploring Local Cultures, Global Resources, and Regional Identities," *Environment and Planning D: Society and Space* 14, 273-92.

—— (1997), "Popular Music in a Transnational World: the Construction of Local Identities in Singapore," *Asia Pacific Viewpoint* 38:1, 19-36.

Krims, A. (2000), *Rap Music and the Poetics of Identity* (Cambridge: University Press).

Kruse, H. (1993), "Subcultural Identity in Alternative Music Culture," *Popular Music* 12:1, 33-41.

Leonard, M. (1998), "Paper Planes: Travelling the New Grrrl Geographies," in T. Skeleton and G. Valentine (eds.).

Leyshon, A., Matless, D., and Revill, G. (eds.) (1998), *The Place of Music* (New York: The Guilford Press).

Lockard, C.A. (1998), *Dance of Life: Popular Music and Politics in Southeast Asia* (Honolulu: University of Hawai'i Press).

Maffesoli, M. (1996), *The Time of the Tribes: Decline of Individualism in Mass Society*, transl. (London: Sage).

Mattar, Y, (2003), "Virtual Communities and Hip-hop Music Consumers in Singapore: Interplaying Global, Local, and Subcultural Identities," *Leisure Studies* 22, 283-300.

Mitchell, T. (1996), *Popular Music and Local Identity* (London: Leicester University Press).

Peterson R. and Bennett, A. (2004), *Music Scenes: Local, Translocal, and Virtual* (Nashville: Vanderbilt University Press).

Power, D. (ed.) (2003), *Profiting from Sound: A Systems Approach to the Dynamics of the Nordic Music Industry* (Oslo, Norway: Nordic Industrial Fund).

Relph, E.C. (1976), *Place and Placelessness* (Research in Planning and Design) (London: Pion).

Rolling Stone (2008), "Best Scene: Baltimore," 1 May, 2008, p. 62.

Scaggs, A. (2008), "The Black Keys' Cure for the Rust Belt Blues," *Rolling Stone* 17 April, 2008, p. 18.

Skeleton, T. and Valentine, G. (eds.) (1998), *Cool Places: Geographies of Youth Cultures* (New York: Routledge).

Scott, A. (2001), "Capitalism, Cities, and the Production of Symbolic Form," *Transactions of the Institute of British Geographers* 26, 11-23.

Straw, W. (1991), "Systems of Articulation, Logics of Change: Communities and Scenes in Popular Music," *Cultural Studies* 5, 368-88.

Two Dollar Pistols. (2004), Personal communication, 9 January 2004, Johnstown, PA.

Wyndham, M. and Read, P. (2003), "Buena Vista Social Club: Local Meets Global and Lives Happily Ever After," *Cultural Geographies* 10, 498-503.

Part VI

The Geography of Genres

The final chapters consider the emergence of three different music genres. In Steven Graves chapter "Hip Hop: A Postmodern Folk Music" the attention is turned to hip hop and rap. He returns to the origin of hip hop in the Bronx in the 1970s to unearth so far unexplored aspects of the development of the genre. Hip hop is imbued with spatial meaning so it is appropriate that geographers approach the genre, something that has scarcely been done in the past. Perhaps some of the most succinct treatments of hip hop's spatiality have been made by non-geographers (e.g. Forman 2002; Krims 2000).

The geographic context of hip hop is perhaps more meaningful than in most forms of popular music. Virtually all rappers carve out an identity for themselves that is inseparable from the spaces that they claim as their territory or a territory they belong to. This has created real or imaginary turf wars within the genre; most notably the infamous East Coast-West Coast divide in the United States. This connection to place, the neighborhood where the artists presumably spent their formative years, and space, the 'hood in a metaphorical sense, offers credibility to their artistic expression. The audience of hip hop is also sensitive to these spatial signals of "authenticity" and understands the music and lyrics based on the perceptions they hold of different places. As Graves discusses, the imagery of violence and drugs that is put forth shows, at best, a partial reality. That image is often driven by the expectations of the audience, and the commercial potential that exists based on these expectations, as much as an artistic need for expression. The concern for identity is not just a dialectic between performer and audience, but can affect the image of cites both discursively and in concrete material ways. Perhaps the most poignant example is the southern California city of Compton southeast of downtown Los Angeles, which has felt the gang stigma emanate from the city's portrayals in hip hop music. City officials have even cited the music as having a negative effect on attracting outside investment and capital (Burbank 2006; del Barco 2006)

The terms hip hop and rap are sometimes used synonymously although a distinction can be made that positions hip hop as a broader culture in which rap exists as its dominant expression. Graves argues that most existing definitions are inadequate and a better understanding of the difference is needed based on the historical development of the genre. He, therefore, presents an interpretation of the historical trajectory of hip hop and rap that should enable music scholars to understand the genre by using traditional concepts, such as popular music and folk music. He concludes that the origin of hip hop exhibits strong similarities with other musical forms that made a transition from folk to popular music. While the hybridity of expression, the self-referential nature of the genre, and the frequent borrowing and collage characteristics of hip hop and rap makes it a "postmodern" music, Graves makes an argument for the existence of a postmodern folk music. As a contribution to the geographical understanding of music in general, Graves also argues that folk music should be defined as music by and for a local audience, unlike popular music which, by its very nature, exists on a broader geographic scale. Hip hop's transition to rap can illuminate this distinction and also broaden the styles of music that we normally consider "folk."

The step from hip hop to techno, which is the genre under investigation in Deborah Che's chapter "Techno: Music and Entrepreneurship in Post-Fordist Detroit" is not very big for several reasons. First of all, techno is also a form of DJ-based music that originated within an urban African American community. There is also an overlap in the temporal development of the genres, although techno emerged locally a few years later than hip hop. This lag resulted in the global spread and popularity of techno music occurring slightly later. Che's treatment of techno is also similar to Graves approach to hip hop as the chapter emphasizes a historical interpretation of the musical genre. Both authors have identified significant gaps in the existing knowledge about the emergence of these genres, which justifies revisiting music that previously has been discussed in the literature (Sicko 1999; Reynolds 1998).

While techno, much like hip hop, is strongly connected to the environment in which it developed, the context is different, which has led Che to use a very different conceptual framework to shed light on techno, its place-based character, and how it is situated in a changing world. Che interprets techno as connected to the transition from Fordism to post-Fordism. As these concepts are well known to economic and urban geographers, the discipline of geography is uniquely positioned to apply them to new fields of inquiry, in this case music, and Che has seized the opportunity. In perhaps a less overt way than Lindenbaum's subsequent exposé of the Christian music industry, but no less fruitful, Che is therefore establishing much needed connections between economic geography and music geography.[1]

In Che's analysis, the techno pioneers' use of new technology in an inexpensive, yet inventive do-it-yourself fashion, becomes an entrepreneurial post-Fordist endeavor rather than a folk music as Graves interpreted hip hop to be in the Bronx. The two ways of understanding an emerging musical form are not mutually exclusive. The different geographies discussed here are, however, the reasons why it is appropriate to interpret hip hop and techno differently. While Graves view the Bronx as a relatively isolated "urban island," Detroit circa 1980 is in the midst of dramatic economic changes, which is ironic because it was, of course, the home of the Fordist production system in the first place. To further ground the musical development of Detroit in the city's socio-technical relations, Che

---

1   The work of Andrew Leyshon and others (e.g. Leyshon et al. 2005) has also established paths towards new research on this important aspect of music geography.

portrays the "Motown sound" as Fordist in contrast to post-Fordist techno. While music from different eras in Detroit are not always musically connected (e.g. Iggy Pop, Eminem, the White Stripes), the continuity is nevertheless evident. This continuity shows that long-term trends that favor musical innovation are place-based (as previously discussed in Chapter 13 by Johansson and Bell).

Techno was inspired both by the built industrial environment of Detroit and its sonic landscapes. In its method of production, techno thrived on new technology and the opportunities that technology presented. Early techno in Detroit is a classic case of an industry in the early stages of the production cycle with low barriers to entry and rapid innovation. But unlike hip hop's transformation to the global cultural force that is rap today, techno has remained a niche genre that appeals to a enthusiastic, but smaller segment of the popular music market. What Che's investigation of techno also shows is the global connectedness of popular music, where influences move globally, in this case primarily between Europe and Detroit, back and forth to continuously create new musical expressions. Techno today is a segment of the broader wave of electronic dance music that it helped to spawn.

The emphasis of techno is on quick beats suitable for dancing and parties. While politico-economic trends, as Che explains, were instrumental to the development of techno, the music itself is not explicitly political. The middle class youth in Detroit that created techno were privileged enough to view technological changes in a positive light, seeing the opportunity rather than just the destruction that post-Fordism wrought on Detroit, and, they channeled that technological change through their music. Interestingly, this is a very different and more optimistic take on deindustrialization and Western society's economic metamorphosis than what is expressed in other forms of music of the time, such as the British punk rock movement.

Compared to techno and hip hop, contemporary Christian music is a genre that is, for the most part, unexplored by the music research community. At least we can say that geographic appraisals of the genre are non-existent. Perhaps the genre holds less personal appeal for many music geographers to be taken seriously as a research topic. Whatever the reasons are, John Lindenbaum's "The Production of Contemporary Christian Music: A Geographic Perspective" fills a void. The fundamental geographic aspects of contemporary Christian music have hitherto not been mapped. In his chapter, we catch a glimpse of these features. Although Lindenbaum traces the roots of contemporary Christian music to southern California, the emphasis of the chapter is on space and the ubiquitous nature of the genre in United States today, rather than a particular place. Contemporary Christian music is not consumed to an equal extent across United States, but as Lindenbaum shows, neither is it limited to areas—rural America, the Bible Belt— that in the popular imagination are most likely to be consumers of the genre. While Lindenbaum maps consumption features such as album sales and summer music festivals, a salient point to make about contemporary Christian music is that the production of the music is not in the hands of enthusiastic amateurs working with a missionary zeal as a prime motivation; rather, a large profit-seeking industry has developed around contemporary Christian music. And as a big business, it has also developed an economic geography that is shaped by the spatial strategies of large corporations. A handful of gatekeepers of contemporary Christian music in media and retail also practice acts of delineation to control the industry. The industry is heavily concentrated in Nashville, Tennessee, not only because the required industry infrastructure can be found there, but also because the markets, demographically and geographically, for contemporary Christian music overlaps with the other musical style associated with Nashville, country music. This makes contemporary Christian music a natural fit for the production and marketing skills that have developed in Nashville. Lindenbaum's emphasis on economic-geographic themes, including ownership patterns of labels, distributors, Christian radio, and so on, is a welcome addition to this volume where the economic perspective does not occupy much space; however, Lindenbaum's chapter, together with Deborah Che's treatment of post-Fordist Detroit, point towards opportunities for further use of theories and methods from economic geography in the geographic analysis of popular music.

References

Burbank, L. (2006), Terror, Hope on the Streets of Compton, Part 1 and 2 [National Public Radio soundfiles] <http://news.nwpr.org/templates/story/story.php?storyId=5247323> and <http://news.nwpr.org/templates/story/story.php?storyId=5248673>, accessed April 2007.

del Barco, M. (2006), A Deadly Year for the City of Compton [National Public Radio soundfile]. <http://news.nwpr.org/templates/story/story.php?storyId=4804296>, accessed April 2007.

Forman, M. (2002), *The 'Hood Comes First: Race Space and Place in Rap and Hip Hop* (Middletown, CT: Wesleyan University Press).

Krims, A. (2000), *Rap Music and the Poetics of Identity* (Cambridge: University Press).

Leyshon, A., Webb, P., French, S. and Crewe, L. (2005), "On the Reproduction of the Musical Economy After the Internet," *Media, Culture & Society* 27: 2, 177-209.

Reynolds, S. (1998), *Generation Ecstasy: Into the World of Techno and Rave Culture* (New York: Routledge).

Sicko, D. (1999), *Techno Rebels: The Renegades of Electronic Funk* (New York: Billboard Books).

Chapter 14

Hip Hop: A Postmodern Folk Music

Steven Graves

Rap music has taken its place alongside pop, rock and country music as a multi-billion dollar cultural industry and a significant cultural force worldwide. This popular and controversial genre has invited ample analysis from media pundits and academics alike, though little has come from geographers. This is unfortunate not only because rap musicians and fans are especially spatially aware, but because a proper foregrounding of the spatial elements in the evolutionary trajectory of rap highlights instructive similarities between it and other popular music genres. Like the blues, country and rock music, rap also evolved from a folk antecedent performed by amateur musicians for a local audience, largely without promise of significant monetary gain. This folk antecedent was called "hip hop" during the 1970s, when it could only be experienced in the Bronx. The terms "rap" and "hip hop" are now used interchangeably, but in fact, I would argue that they are quite distinct musical forms and a clarification of terms is long overdue. Rap music is a popular art form. It is global, commercial and focused on the MC or rapper. Hip hop on the other hand was a folk art that exhibited many traits found in other music genres widely acknowledged as "folk music".[1] Among these traits were its adherence to oral traditions long established in the African-American musical culture, a home grown DIY (do-it-yourself) creative ethic, and the amateur status of its artists. The most critical folk characteristic of hip hop was, however, its strictly local orientation. Much of the remainder of this chapter is devoted to defending the assertions here by exploring the history of hip hop and the evolutionary changes wrought to it upon its diffusion from the Bronx.

The efficacy of spatial constructs as a tool for demarcating boundaries between rap and hip hop invites a reconsideration of the way in which popular culture in general is differentiated from folk culture. A secondary thesis in this chapter posits that space is the fundamental element defining both folk culture and popular culture. Such a reformulation is worth consideration, because aspatial conceptualizations of "folk culture" and "popular culture" are quite common across the academy, even finding favor in many of the more popular introductory geography textbooks (see e.g. Rubenstein 2008; Knox and Marston 2007). Geographers Connell and

---

1 Hip hop flourished in the Bronx from 1973 to the early 1980s, and for this reason it is generally referred to in this chapter in the past tense. It is acknowledged later in this chapter that hip hop is still played/practiced in a form similar to the variety performed in the Bronx in the 1970s.

Gibson (2003) thoughtfully engage the problems of distinguishing folk music and popular music, but they too appear to eschew spatialized definition of popular when they defer to Grossberg (1997) who rejects (or possibly fails to even consider) the utility of spatial criteria in defining the "popular." There is good reason to follow Grossberg's lead because to do otherwise requires wading into highly contentious waters surrounding questions of authenticity, credibility, and aesthetic intention. Connell and Gibson conclude, "There can be no formal definition of popular music" (2003, 5). This may be true in the strictest sense, especially at the fuzzy boundaries between folk and popular arts, but it also seems an unnecessary concession. In most instances, folk arts can be distinguished from popular arts by referring to the spatial relationship between artists and audiences.

Another reason why it is so difficult to think of hip hop or rap music as folk art is that it has been widely labeled as the first "postmodern music" (Shusterman 1991; West 1988). This is partially an outgrowth of the unusual degree of triumphant and unwarranted exceptionalism attributed to hip hop and rap. In addition to the hyperbole, hip hop and rap indeed exhibit multiple postmodern elements. The abundant use of sampling and a loose adherence to traditional notions of musical form and structure is the most obvious postmodern characteristic of hip hop and rap. By incorporating snippets of others music, without regard to genre or stylistic convention, into a new pastiche, hip hop is parallels almost perfectly the techniques used by postmodern architects who slap together disparate styles into a coherent whole. Hip hop and rap music also make ample use of irony, another common marker of postmodernity. Both genres often use the music of other composers in a fashion at odds with the original composer's intent and to audiences far removed from the original (see Ogg 2001). In addition to the obvious pastiche characteristics that marked early hip hop, its pop incarnation (rap) exhibits yet another common characteristic of postmodernity—its appeal to street authenticity has largely been constructed to appeal to the public's imagined *ideal* of black inner city life, while dismissing honest representations of the inner city. Hip hop produced in the South Bronx during the mid-1970s (the quintessential inner city time/place) is now considered too soft or cartoonish to authentically represent inner city street culture by those *who have never lived or experienced the inner city*. The commodification of a fetishized narrative of inner city street culture resulted in the rejection of the South Bronx's home grown narrative in favor of one constructed about inner city life by middle class artists and their production teams. While these observations are by themselves compelling, they also serve to illuminate rap's evolutionary trajectory from a local folk art, which is the focus of the following paragraphs.

The evolution of hip hop

Hip hop evolved rapidly in the Bronx during the 1970s as one component of a host of youth activities that included break dancing and graffiti art. During this era, the Bronx was widely considered the epicenter of urban decline in America. Gangs were a major problem during the early 1970s, but began to decline in importance as hip hop rose to prominence. Hip hop's role in reducing that violence is debatable, but does appear to have provided several non-lethal, ritualistic battle grounds (see Caz in Kugelberg 2007).[2]

By all accounts, the first hip hop "jam" (performance) was a back-to-school dance DJ'ed by Kool Herc (Clive Campbell) in 1973 at 1520 Sedgewick Avenue, north of Yankee Stadium. It was reputedly just a fundraiser thrown by Herc's sister who wanted to buy some new back-to-school clothes (Chang 2005). Because many of the seminal figures in hip hop attended this initial performance, and it provided a template upon which hip hop as a performance genre evolved. Probably no single element of that first performance was revolutionary in its own right, but combined as they were by Herc, they constituted the birth of a new musical genre. Had this first hip hop jam simply featured Herc playing disco records, it would have been little different than thousands of school dances around the US and solidly within the realm of popular culture practice, but this dance featured a number of practices that mark it as a unique folk culture. Among the elements that made this dance party different were: 1) the DJ *made* music, instead of just playing the music of others, blurring the line of "ownership" of the music; 2) an MC or rapper was employed to enhance the DJs performance 3) the creativity of the DJ and MC established a reputation that functioned as a currency more important than the monetary reward derived from the audience; and 4) most importantly, the performance was customized to suit the taste preferences of a very local audience, whose taste preferences diverged considerably from national or even regional taste preferences for dance music. A discussion of these and other folk-like elements of hip hop follows.

Folk characteristics

Oral tradition

The orality of hip hop is the most easily recognized folk-like characteristics of hip hop. The lack of a organized, formal, text-based means of transmitting the cultural traditions within hip hop place it squarely in line with other folk traditions that rely upon word-of-mouth for codification. This oral tradition is common in other black music genres. Rapping or MCing stands at the long end of a line of oral traditions, that some claim extends all the way back to Africa (Toop 1984; Perkins 1996). Clearly there are more contemporary antecedents. Jive talking radio disc jockeys, prison toasts, a variety of spoken word albums, the competitive insult game "the

---

2   Because stage names are both commonplace and far more recognizable than performer's given names in the hip hop idiom, stage names are herein used in both text and references.

dozens" and even Muhammad Ali have been cited as inspirations to the early MCs (Perkins 1996, Chang 2005; Kugelberg 2007).

Though it is hard to prove, the oral traditions in hip hop may have been most directly descendent from Jamaican antecedents (e.g. Hebdige 1987). Others suggest convincingly that the connections between African, Jamaican and Bronx traditions are tenuous at best (see Ogg 2001). Kool Herc was a Jamaican immigrant who was well aware of both DJ and MC practices in his homeland and helped introduce variants of both into the Bronx, which had a culture predisposed to incorporating these innovations. Hip hop appears to borrow some oral elements from Jamaican "yard culture," where "boasting and toasting" is a common means to display wit and gain prestige (Perkins 1996). Hip hop also seems to borrow much from Jamaican dub music, a genre invented when the vocal track was accidentally left off a record. These vocal-less recordings became favored by both dancers and DJs for their focus on the danceable rhythms, and the space they permit the MCs for vocal improvisation (Chang 2005).

Though Kool Herc did not use dub records as part of his performance, he replicated the dub formula by selecting for use specific instrumental passages on songs to please his audience. He also used an MC (rapper Coke La Rock) to call out improvised rhymes to the dancing audience. The MC served both to energize the crowd and to boast about the skill of the DJ. The role of the MC grew and the ability to boast about oneself, the DJ and the party became increasingly important, evolving into a competitive art form in which the ability to also denigrate other MCs also became valued. Despite the lack of formal training, the oral component of hip hop became surprisingly sophisticated, while remaining somewhat improvisational and unwritten (e.g. Kugelberg 2007).

The competitive nature of rapping parallels the evolution of competitive break dancing, DJing and graffiti art. Each served as one alternative to gang violence by offering opportunities for the establishment of individual identity and group membership. The MC, in later years becomes more prominent, and eventually the focal point of the commercialized rap music.

DIY – homemade creativity

Another element of hip hop marking it as a folk art is the improvisational, do-it-yourself art ethic that characterized the DJs performances. Kool Herc again is among the first to debut this folk-like practice (see Jaquan.com 2006). No doubt conscious of the formula that drove forth the success of dub music in Jamaica, Herc had to adapt it to the Bronx, because local audiences did not respond to Jamaican dance music. Noting that dancers responded more enthusiastically to specific percussion-heavy segments of songs, Herc introduced a technique he called the "Merry Go Round" to extend these drum "breaks" (Ogg 2001; Chang in Kugelberg 2007). By using two of the same record on separate turntables, Herc found that he could lengthen a desired song segment by playing that segment on a second turntable, just as it ended on the first. By moving the needle on the record back to the beginning of that segment on the first turntable and repeating the process, he could extend drum breaks indefinitely. Because Herc DJ style focused on playing drum breaks, he dubbed those dancers who danced mostly to these parts, "break boys," later shortened to "b-boys (and girls)."

By manipulating records as he did, Herc transformed the role of disc jockey from a passive medium for music created by others, into an active participant in the composition of new music. The fact that this creative process emerged in the South Bronx in the 1970s seems logical given how costly making music in a standard multi-piece band must have seemed to many in Herc's audience, especially since many of them were pre-teens in 1973. Similar conditions may have driven the evolution of Jamaican DJ culture (see Chang 2005). Still many aspiring DJs noted Herc's success with this formula and quickly set out to refine and extend what he had done.

Several key innovations were introduced into the hip hop musical idiom in the next couple of years that significantly changed the role of DJ. In 1975, DJ Grandmaster Flash (Joseph Sadler), using a two-turntable system of his own design and construction, began to mix increasingly smaller segments of music into a new collage or mix, that could legitimately considered a new song in its own right. Flash had learned part of this technique from Herc and partly from another DJ friend, Pete Jones. Flash perfected the technique through jerry-rigging his equipment and making careful observations of the record as it moved on the turntable and using both to perfect his techniques. This was so revolutionary, that this new technique backfired at first because his audience was so captivated by his DJ'ing style that they failed to dance as he had hoped (Chang 2005).

The final major musical innovation of the hip hop DJ that established the DJ-as-musician was a technique called scratching. Scratching, or moving the record back and forth while the needle is still in the record was first played for a crowd by Flash's friend, Theodore Livingstone who debuted the technique in 1975 (Kugelberg 2007). He was twelve at the time. Scratching further transformed the turntable into a rhythmic instrument, rendered the recorded music on the vinyl unimportant and unrecognizable. The DJ was now a musician. This innovation in turn led to several other smaller innovations that cemented the role of DJ-as-musician (Kugelberg 2007). Eventually, the practice of making music using these techniques became known as "turntablism" and the DJ as "turntablist" (Price 2006).

Paralleling the homemade nature of the music was a DIY attitude regarding the construction of the musical instruments that also marks hip hop as folk music. Because hip hop artists were frequently without resources to ply their craft, they were forced to make do with locally available materials. This practice recalls the often repeated stories of folk-blues guitarists (including an early B.B. King and Bo Diddley) fashioning homemade guitars from scrap wood and spare wire. Several DJs tell stories about how they taught themselves how to construct sound systems and DJ equipment (e.g. Chang 2005; Toop 1991). Flash was particularly adept at electronics, constructing several of the mechanisms necessary for him to create his musical montages. Because hip hop audiences were frequently under 18, and may not have been able to meet the dress code (no sneakers) required by local discos (see Caz in Kugelberg 2007; Ogg 2001), hip hop jams were frequently held out of doors in parks, where the electricity needed to power the sound systems was obtained by wiring directly into municipal lamp posts or public buildings (Chang 2005). Kugleberg (2007) calls the culture "self-starting." Indeed hip hop was homemade. Much has been made of DIY culture among punk rock subculture, but hip hop was in many ways far more DIY (Ogg 2001).

A non-commercial art

One of the distinct markers of folk music is its lack of a clear profit motive among performers and fuzzy notions surrounding the "ownership" of songs. Appalachian folk musicians have been identified by their amateur status and a tendency to stick to a common musical repertoire, endlessly tweaked by performers. Hip hop DJs and MCs behaved in much the same manner in the Bronx during the 1970s. Clearly, there was little regard for the legalistic notion of copyright in early hip hop, with DJs as a matter of course taking bits and pieces others music to make something that they might claim as their own (Kugelberg 2007). Delta Bluesmen and Jamaican musicians also have long histories of this attitude. Jamaican musicians are well known for their habit of repeatedly covering popular songs. Known as version in Jamaica, a single recording may have dozens versions, rerecorded by competing artists simultaneously available to the record buying public. An example cited by Dick Hebdige is Wayne Smith's "Under Mi Sleng Teeng," released in 1985 which had no less than 239 versions released in its wake (1987). So engrained in the culture was this attitude that it carried on for some years after hip hop became commercialized rap. It took a series of high profile lawsuits against rap musicians who had liberally sampled the music of others before a formal recognition of musical ownership was established. Eliminating sampling as it was done in the Bronx represents another structural change in the music when it became rap.

Although it is nearly inconceivable today, when hip hop was still a local music for local audiences, performers and audiences alike were apparently unaware that the music had significant market value. Grandmaster Caz, an early MC, found the notion of professionalism silly, noting "… [hip hop] hadn't spread wide enough for people to think that it would catch on. It was like something that kids do" (Kugelberg 2007, 200). Admittedly, modest sums were generated by DJs who struggled to make a living playing parties and doing club gigs. Most had some other form of income, and many lived at home with parents. Of the hundreds of hip hop DJs in the region, only Kool Herc appears to have been capable of fashioning a full-time job as a club DJ. Others arranged dance parties in discos and other venues, especially during the winter months, but until the 1980s, hip hop remained almost exclusively an art form for amateurs, highly skilled as they may have become (see Caz in Kugelberg 2007). By the mid-1970s, crowds of thousands were attending hip hop performances in parks, but gate receipts were small at best. Modest sums were also made from the sale of taped recordings of the shows. DJ Grandwizard Theodore noted, "We didn't really do it for the money. We did it because we loved it. The average old school MC or DJ or whatever, they got into it because they loved it" (quoted in Ogg, 2001, 34).

Reputation was the valued currency earned by becoming an accomplished DJ; and among DJs, three characteristics were valued above all others. The first marker of a good DJ was the quality and volume of his sound system. DJs who had acquired good equipment and mastered the technical obstacles associated with producing massive volume were held in high esteem. Herc, as was the case with DJs in Jamaica, had a powerful, acoustically superior sound system (see Chang 2005; Ogg 2001). This characteristic emerged in other folk genres that also had a significant dance element while constrained by repeated performances in makeshift, acoustically challenging locations (barns, parks, and so on). These genres value the primacy of volume and minimize the role of traditional (e.g. singing) vocal accompaniment.

Secondly, good DJs had to be knowledgeable music librarians, with large collections. Whereas musicians in the traditional sense have often been the children of musicians themselves, hip hop DJs were more often from families with eclectic musical tastes and large record collections. Herc's father had a large collection of records, including a significant number of country music albums (see Chang 2005). In order to protect his reputation Herc tried to obscure the identity of records that audiences loved. Herc would wash the paper labels off records to keep rival DJs from stealing his secrets (see Caz in Kugelberg 2007). The master of music librarianship appears to have been DJ Afrika Bambatta, another key figure in the development of hip hop, who began DJing at the age of 10. His record collection was immense and he vastly expanded the repertoire of available breakbeats used by DJs. Most DJs relied upon a handful of well known breakbeats to build their performances, including several James Brown tunes, a few disco songs and several tunes by the Incredible Bongo Band. Bambatta introduced a number of rock songs, new wave and punk sounds into the hip hop repertoire.

The third valued asset of top hip hop DJs was the technical virtuosity of the DJ. Disc jockeys like Flash and Theodore were dubbed "Grandmaster" (a reference to Kung Fu movies of the era) because of the speed and skill with which they could scratch and mix many different snippets of music. DJs who had mastered these physical challenges, would emphasize their ability by scratching with their elbows, facing away from the turntables and incorporating other acrobatic feats into their performance, much like a guitar player might do with his instrument.

A local art for a local audience

The last folk-like characteristic of hip hop, and the one most often ignored by scholars of culture, was its small geographic range. For nearly 10 years, hip hop was a folk practice confined largely to the Bronx. By the time it began attracting converts in nearby neighborhoods, multiple insiders report that hip hop had begun to fade in its hearth area (e.g. Chang 2005; Ogg 2001).

One of the most remarkable elements of hip hop is the manner in which it demonstrated how cultural and social isolation can replicate the effects of physical isolation in the evolution of cultural practices. In the Bronx, we find a neighborhood so culturally isolated and unique that its residents managed to develop and sustain taste preferences largely in defiance of the considerable marketing machinery headquartered in nearby Manhattan. Members of the hip hop folk culture demonstrated an impressively eclectic musical palate, favoring music that was well out of step with that played on Top 40 radio (pervasive and nationally uniform during this period) or in disco. Hip hoppers even eschewed much of the music played on 1970s black radio (see Chang 2005).

The apparent lack of outside influences is strong evidence that hip hop was a local music for a local audience. It was a neighborhood's music. Only once it had become rap was it labeled "black" music. Many of those who contributed to the development of hip hop were not African-American. Local Puerto Rican and Dominicans contributed significantly to the musical lexicon of many of the early hip hop DJs. Not only were several influential DJs Hispanic (e.g. DJ Charlie Chase), but more importantly a significant component of the local DJs audience (including b-boys/girls) were likely to be Hispanic and therefore able to exert influence on the musical direction of DJs (see Del Barco; Flores in Perkins 1996 for an elaboration of Hispanic influences).

Space has remained an important issue in hip hop and rap music. Rap fans are frequently highly conscious of the hometown of favorite performers. Proper urban credentials are critical in the marketing of these acts. This has been the case for many years. Consider for example that it took nearly 10 years before a rap act from beyond Greater New York City to appear on the Billboard charts, and even then it was a sort of "kiddie rap" (DJ Jazzy Jeff and the Fresh Prince). White rappers whose legitimacy and/or authenticity are critical to their acceptance, even by white audiences, have been judged in part by their geographic credentials. The Brooklyn based white rap group 3rd Bass were early entrants into the hip hop mainstream were accepted by black rap audiences in part because of their faithfulness to the idiom, but also because they were from an acceptable neighborhood. The Beastie Boys, a popular white rap act, were less favorably received by "purists" in part because although they were New Yorkers, were clearly from the wrong neighborhoods. Special venom was reserved for the popular white rapper, Vanilla Ice, who successfully deceived audiences for some time by claiming that he was a product of Miami's gang culture. He was in fact from suburban Dallas. The most popular white rapper, Eminem was from a tough area of Detroit and so has had an easier claim to authenticity.

Global transformation

In 1979, just as hip hop was dying out in the Bronx, "Rapper's Delight" was recorded by the Sugar Hill Gang, thus beginning the rap era. Hip hop, the DJ-centered, locally produced and consumed, performance folk art rapidly descended into obscurity, displaced almost wholly in a few years by an MC-centered, globally produced and consumed, medium constricted popular music genre.

The very early history of rap music serves to highlight both the folk characteristics of hip hop and the truly radical departure from hip hop that rap music quickly becomes. The details surrounding the recording of "Rapper's Delight" vary, but most accounts agree that the members of the Sugar Hill gang were not among the Bronx's most accomplished MCs. Instead, the trio of rappers was hastily assembled by Sylvia Robinson, an executive with the small New Jersey-based independent record label, Sugar Hill Records. Robinson was exposed to the Bronx sound via cassette tape recordings of hip hop performances. Hundreds of these tapes were in circulation around greater New York and were popular especially with aspiring MCs who listened to the tapes to learn how to rap (Perkins 1996; Caz, in Kugelberg 2007). Robinson heard some kids practicing their raps and took a chance on recording three aspiring MCs in the summer of 1979. Several other rap records also were recorded at about the same time, notably Fatback's "King Tim III (Personality Jock)." Notably, all of the early rap records were produced by acts only tangentially involved in the Bronx's hip hop culture.

It is interesting to note the numerous parallels between the trajectory of early rap and early rock n' roll, especially the similarities between the Sugar Hill Gang and rock 'n' roll pioneer Bill Haley and the Comets. Like Bill Haley's "Rock Around the Clock," "Rapper's Delight" was in many ways a pale imitation of the musical genre it mimicked. Both songs were essentially novelty records, recorded as experiments. The shortcomings of the Sugar Hill Gang did not dampen the enthusiasm for their record outside of the Bronx. The rest of the world was not familiar with the sound of the original, and may have not liked the Bronx style hip hop anyway. "Rapper's Delight" rose into the national pop and R&B charts, and became a staple of urban radio in 1980, especially in New York City.

Within the Bronx's hip hop community "Rapper's Delight" was met with both derision and confusion. The long rambling lyrics to "Rapper's Delight" were concocted by Bronx based hip hop artist Grandmaster Caz (a neighbor of Herc) and were shared by Caz with "Big Bank Hank," a friend who worked as a club bouncer and pizza chef. Big Bank Hank was recruited into the Sugar Hill Gang and used Caz's rhymes as his own. Because the members of the Sugar Hill Gang were not truly members of this folk culture, they were able to "steal" from it, without suffering the consequences of breaking the rules of fair competition observed within the Bronx.

Caz was surprised that a demand for his raps extended beyond the narrow geographic confines of the ghettos of New York. Caz was shocked when he heard his lyrics on the radio. Caz recalled, "... every car that passed by had

['Rapper's Delight'] on the radio. Everybody knew those rhymes were mine and half were coming up to me, 'Yo, I heard you on the radio!'" (quoted in Hager 1984, 50). Though Caz was never compensated for his contribution, he and others seemed most angry that a performance that would not pass muster in their own neighborhood had become a hit, and that his hard won reputation had been stolen (see Ogg 2001; Kugelberg 2007)

The most important break with the folk traditions established in the Bronx was the transformation from live performance art to recorded music. This transformation was necessitated by the requirements of mass distribution via vinyl records. A hip hop performance could not be put on a record. A typical hip hop performance was largely focused on the performance of the DJ, required active audience participation, and could last many hours. Putting all this on a record required a transformation that might be likened to the reduction of an entire ballet or an opera to a radio-friendly, three minute 45 rpm record. It should be noted that "Rapper's Delight" captured, at least in part, some of the ambition of the original art, by running some 15 minutes in length in total disregard of the rules governing Top 40 radio play. Nevertheless, it did get considerable airtime on many urban-format radio stations, especially on the East Coast. That this gamble paid off was surprising to many in the mainstream who considered "Rapper's Delight" way too long for radio, but it was even more surprising to hip hop insiders who couldn't believe that a rap performance had been abridged to fit onto a record (see Chuck D's reaction in Chang 2005; see also Kugelberg 2007). Also, insiders were surprised that very few of the early rap records used a DJ. Traditional bands, who do little more than play a repetitive background riff for the rappers replaced the DJ in the studio.

Few of the Bronx based hip hop acts were interested in making records initially. Kings of the Bronx scene, Grandmaster Flash and his crew the Furious Five were early targets for the entrepreneurial indie label bosses, but Flash had no interest in a record deal. Chang notes, "To [Flash], the idea was absurd. Who would want to buy a record of Bronx kids rapping over a record?" (quoted in Chang 2005, 129) To them, playing for a live audience was a safer bet, provided some income and preserved hard won reputations. Eventually though, it was the preservation of reputation that forced many hip hop artists into making a record when a record deal became a status marker (see Kugelberg 2007).

When DJs, like Grandmaster Flash did make a record, they didn't appear themselves on their own records, working instead as quasi-producers who helped craft what the bands played in the studio. Flash is not even on the "his" early single "Superrappin" record, even though his MC crew, the Furious Five, boast in the song of Flash's mastery of the turntable and beatboxes. Afrika Bambaata's first commercial release "Planet Rock" was also a significant departure from the live performance style that he had helped pioneer.

In addition to the changes necessitated by the change in media, the lyrical content, presentation style and politics of hip hop were changed by rap artists and their producers in order to sell records to audiences outside the Bronx. The transformation was swift. Abandoned first was the upbeat dance-party lyrical themes that marked hip hop as a healthy escape from the gang violence and the indignities of ghetto life. Although not G rated, hip hop raps were believed (perhaps correctly) to be too soft and upbeat to appeal to those seeking "authentic" voices from the inner city. Middle class audiences, who later proved to be the most lucrative audience for rap, wanted a darker, tougher vision of ghetto life. Rappers and clever production teams were only too happy to oblige. The first rap group to successfully use this marketing strategy in order to appeal to national, mostly white audience was the Hollis Queens-based group Run-DMC. Deftly produced by a middle class, Jewish, college student from Long Island, Run-DMC presented themselves as authentic street-toughs, though in reality none had been in a gang and all came from stable, middle class families. Run-DMC even featured hard rock guitar riffs in a number of their earliest works. Kurtis Blow was the first rapper to be signed to a major label deal. He was a middle class college graduate from Harlem, and an associate of the rap impresario-to-be Russell Simmons, one of the chief architects of the hip hop-to-rap evolution. Insider Grandmaster Caz called him a "disco rapper" (Kugelberg 2007).

When Bronx based hip hop acts tried their hand at making records they too largely followed the same marketing strategy. David Samuels's oft-anthologized essay on the evolution of rap contains a revealing story about the reception similar marketing efforts received in the hip hop's hearth place:

> Like disco music and jumpsuits, the social commentaries of early rappers like Grandmaster Flash and Mellie Mel were for the most part transparent attempts to sell records to whites by any means necessary. Songs like "White Lines" (with its anti-drug theme) and "The Message" (about ghetto life) had the desired effect, drawing fulsome praise from white rock critics, raised on the protest ballads of Bob Dylan and Phil Ochs. The reaction on the street was somewhat less favorable. "The Message" is a case in point. "People hated that record," recalls Russell Simmons, president of Def Jam Records. "I remember Junebug, a famous DJ of the time, was playing it up at the Fever, and Ronnie DJ put a pistol to his head and said, 'Take that record off and break it or I'll blow your fucking head off.' The whole club stopped until he broke that record and put it in the garbage" (quoted in Samuels 1991, 25).

Simmons, the brilliant architect of Def Jam records clearly recognized that lyrical content and ghetto imagery of this sort wasn't for local audiences, but was perfectly suited for record buyers in the middle class. Run-DMC rode their concocted street tough imagery to stardom, even taunting their Bronx-based forerunners in some of their earliest records, perhaps in an attempt to position themselves as more authentic representatives of inner city life than actual (former) gang members like Afrika Bambatta. Run-DMC brought rap fully into the mainstream in 1986 when they collaborated with classic rock stalwarts Aerosmith in a cover of their mid-1970s hit "Walk this Way." Later groups that came to represent authenticity and "street cred" were frequently not actually what they appeared to be. Groups like N.W.A. and Public Enemy were largely composed of middle class blacks from suburban areas according to Samuels (1991). When Soundscan (an accurate strategy for measuring record sales and audiences) came online in the spring of 1991, the wider music industry came to realize what people like Simmons had known for some years: the audience for rap music was largely white, male and suburban.

Conclusion: Rethinking definitions

This spatially focused history of hip hop music suggests that several commonly held notions about music genres deserve reconsideration. First, it must be argued that there actually is a difference between hip hop and rap music. Second, the idea that folk arts are defined by their degree of nostalgic appeal must be disposed in favor of a more useable definition that focuses on the spatial contexts of their production and consumption.

Hip hop is not rap music

When rap music first emerged in 1980, it was called "rap music" both by the media and by those who were performing it. Many of the early songs were called "raps." Some recognized the shift this represented immediately. Hip hop documentarian and filmmaker Charlie Ahearn said as much when he proclaimed, "Hip hop is dead by 1980. It's true" (quoted in Chang 2005, 132). Few people used the term hip hop to refer to rap music for around ten years, but in the 1990s, the term hip hop began to reemerge, especially among those who considered themselves knowledgeable about rap music, or at the very least those who wanted to appear knowledgeable about rap music. Industry magazines, such as *The Source* appear to be partly responsible for appropriating term hip hop for rap. This sleight of hand appears to be just one among hundreds of maneuvers by the rap music industry to claim some measure of ghetto authenticity in a highly commercialized, tragically inauthentic cultural media. The original nomenclature was correct and should be restored, though there's little chance of that happening unless the taste makers within rap music decide to do so.

   Latter day scholars of hip hop hint at some of the problems with bundling hip hop and rap. The definitions offered by Price (2006) implicitly acknowledge the differences, but do not recognize the significant difference in the two art forms:

> *Rap:* The music of Hip Hop Culture. Traditionally, rap music consisted of the combined talents and creativity of both an MC and a DJ, but most current rap music does not involve a DJ. This genre of music consists of local and regional styles as well as different philosophical approaches and includes gangsta rap, East Coast rap, West Coast rap, southern rap, and conscious rap, to name a few (323).

Hip Hop: The musical expression of Hip Hop Culture. Although it is often differentiated from rap music, it is similar to rap in its use of DJs, MCs and producers. Usually lowercase (hip hop) (320).

Some have offered the slangy term "Old School" as a way of acknowledging differences between more recent rap styles and the early years of rap and hip hop (see Price 2006). This term is helpful because there is some continuity in the vocal style used in hip hop and rap music through the mid-1980s. However, this designation fails to account for the other significant differences in the DJ-centric performance art of the Bronx, and the MC-centric rap music produced in record studios. A genuine hip hop jam as performed in the Bronx in the 1970s can not be put on a record. Rap songs can. The spatial transformation from hip hop as a local performance for a local audience to a rap song for a wide audience required a fundamental structural change in the art form. The degree of this transformation is significant, and represents a transformation more radical than the changes field hollers and Celtic folk music underwent as they evolved into popular, commercial art forms. Hip hop's transformation into rap is more akin to the reduction of an entire opera or ballet (costumes, stage sets, and so on) to a radio friendly pop song. It is analogous to claiming that a single of the "Toreador Song" is an opera.

The final problem with blurring the distinction between rap music and hip hop performance that it ignores the fact that hip hop still exists, and remains quite different from rap. Hip hop enthusiasts continue to stage performances in the traditional manner, and to suggest that what they do is simply "rap music" ignores the fact that these hip hop "purists" are preserving a broad set folk traditions (e.g. Price 2006).

Space defines folk, popular culture

The other major implication contained in this reconstruction of hip hop's history surrounds the role of space as a defining characteristic of folk culture. First, it may prove instructive to look at how others have defined folk and popular culture. Almost all introductory human geography texts delineate folk and pop culture, and these are important because texts have a wide audience and therefore the great influence on our collective understanding of folk art. Texts also tend to reflect consensus within their home disciplines and the academy in general. This chapter represents an effort to shift the consensus. Rubenstein (2005, 117) defines folk customs as those "practiced primarily by small, homogenous groups living in isolated rural areas…" Popular customs, in contrast, are those practiced in "large, heterogeneous societies that share certain habits," presumably these are urban areas. Rubenstein (ibid.) notes that scale of the territory covered is much smaller than that of a popular culture, but does not elaborate on this key point. Knox and Marston (2007) appreciate the problems with simplistic characterizations of "folk" and "popular" as they apply to culture, but they fail to invoke space as a characteristic capable of defining the differences between popular and folk cultures. A fairly common approach to the question is proffered by Petracca and Sorapure who write,

> folk culture is generally transmitted through oral communication; both [folk and high culture] however, place a high value on tradition, on artifacts produced in the past, and on the shared history of the community … . By contrast, popular culture encompasses the most immediate and contemporary elements in our lives – elements which are often subject to rapid changes in a highly technological world in which people are brought closer and closer by the ubiquitous mass media ... pop culture is the shared knowledge of a specific group at a specific time (1995, 4).

A more contemporary discussion of popular culture is also instructive because it probably best reflects current trends in cultural theory, which also tend to omit spatial considerations. A good example might be how Connell and Gibson (2003) rely upon the influential music scholar Grossberg who writes:

> [the "popular"] cannot be defined by appealing to either an objective aesthetic standard (as if were inherently different from art) nor an objective social standard (as if it were inherently determined by who makes it or for whom it is made). Rather it has to be seen as a sphere in which people struggle over reality and their place in it, a sphere in which people are continuously working with and within already existing relations of power, to make sense of and improve their lives (1997, 2).

This line of reasoning is appealing because it permits Connell and Gibson to suggest that there is no formal definition of popular music (2003), but clearly there are differences in the manner in which music is produced that deserve recognition. This definition also leaves open the possibility that folk cultures by contrast are not within a sphere where people struggle over reality and their place in it.

In general, the role of space has been relegated to an afterthought. The tendency to omit spatial considerations in favor or temporal ones is a well documented phenomena (e.g. Soja 1989), but it is particularly curious when it is done by geographers. The critical role of space in the construction of common, but complex notions such as race and gender, have been advanced quite compellingly (e.g. Mitchell 2002). The logic advanced by geographers like Soja and Mitchell lends itself quite readily to understanding structural (as opposed to stylistic) differences between folk and popular culture and offers an alternative to Grossberg's assertion that the audience and the artist is not a reliable standard for defining folk and popular culture.

Simple and defensible definitions of folk culture and pop culture can be constructed around the spatial arrangement of their audiences. Using this logic, folk music is defined as: "music produced for and consumed by the artist's local community." It is local music for local audiences. Alternatively then, popular music then can be defined as: "music produced for audiences beyond the artist's local community." It is assumed that a "community" is a group of people who interact informally and regularly, and does not necessarily preclude the possibility of cyber communities if they function as a community of people who are familiar with each other.

Using space as the primary defining characteristic of folk art permits a greater focus on the production and consumption of folk arts, while not precluding many of the other defining characteristics offered by scholars of folk and popular culture. A spatial definition offers the possibility of a contemporary, urban folk art; one still likely to be transmitted informally and orally, but one that could be communicated electronically as long as it remained structurally unchanged. If a local art form was altered to appeal to non-local taste preferences, then by definition it would be considered popular music. Spatially constrained folk music is composed without significant promise of monetary gain because their audiences are small by definition, but should artists find that their music suddenly (or not-so-suddenly) gains a profitable global audience, then there would be no need to awkwardly explain why, though unchanged, this art is now pop music.

The spatially inflected definitions of folk and popular arts offered above offer only partial resolution to the sticky issue of authenticity and credibility, which haunt many discussions of popular music. One can never be 100 percent certain of the commercial aspirations of any artist. That point must be conceded. Local musicians may aspire to audiences beyond their community. Pop superstars may conceivably craft music they think will appeal to only their closest friends. If these scenarios do play out, then the form (media) of such performances and their stylistic similarities to prevailing popular taste preferences will in most instances reveal the spatial extent of the intended audience. Music not designed for easy mass marketing via radio or mass production is clearly not pop music. Music that sounds nothing like any other music currently selling well on the market is an improbable candidate for pop music status.

It is worth considering how a conscious change in target audience forces changes in the style, form and structure of art forms. Clearly the shift in attention among hip hop artists away from their own community required fundamental, structural changes in the musical form. Stylistic changes soon followed. This evolutionary trajectory, despite its urban hearth point, mimics the evolution of several other folk genres into popular music. Similar transitions characterize the evolution of the local folk music from the Delta and Appalachia into the Motown sound and Nashville Country.

References

Chang, J. (2005), *Can't Stop Won't Stop: A History of the Hip Hop Generation* (New York: St. Martin's Press).

Connell, J. and Gibson C. (2003), *Sound Tracks: Popular Music, Identity and Place* (New York: Routledge).

Del Barco, M. (1996), "Rap's Latino Sabor," in Perkins (ed.).

Flores, J. (1996), "Puerto Rocks: New York Ricans Stake Their Claim," in Perkins (ed.).

Grossberg, L. (1997), *Dancing in Spite of Myself: Essays on Popular Culture* (Durham, NC: Duke University Press).

Hager, S. (1984), *Hip Hop: The Illustrated History of Break Dancing, Rap Music and Graffiti* (New York: St. Martin's Press).

Hebdige, D. (1987), *Cut 'n' Mix: Culture, Identity and Caribbean Music* (New York: Methuen).

JayQuan (2001), "Pete DJ Jones: Interview Part 1." *The Foundation.* <http://www.jayquan.com/pete.htm>.

Knox, P. (2007), *Human Geography: Places and Regions in Global Context* (Upper Saddle River NJ: Pearson Prentice Hall).

Kugelberg, J. (ed.) (2007), *Born in the Bronx* (New York: Rizzoli).

Marston, S. and Knox P. (2007), *Places and Regions in Global Context: Human Geography,* 4th edition (Upper Saddle River, NJ: Prentice Hall).

Mitchell, D. (2000), *Cultural Geography: A Critical Introduction* (Malden, MA: Blackwell).

Ogg, A. (1999), *The Hip Hop Years: A History of Rap* (New York: Fromm International).

Perkins, W. (ed.) (1996), *Droppin' Science* (Philadelphia: Temple University Press).

Petracca, M. (1995), *Common Culture: Reading and Writing about American Popular Culture* (Upper Saddle River NJ: Prentice Hall).

Price, E. (2006), *Hip Hop Culture* (Santa Barbara, CA: ABC-CLIO).

Rubenstein, J. (2007), *The Cultural Landscape: An Introduction to Human Geography*, 9th edition (Upper Saddle River, NJ: Prentice Hall).

Samuels, D. (1991), The Rap on Rap. *The New Republic.* 205: 24-9.

Shusterman, R. (1991), The Fine Art of Rap. *New Literary History.* 22(3):613-32.

Soja, E. (1989), *Postmodern Geographies: The Reassertion of Space in Critical Social Theory* (New York: Verso).

Toop, D. (1991), *Rap Attack 2: African Rap to Global Hip Hop* (New York: Serpent's Tail).

West, C. (1988), *Prophetic Fragments* (Trenton: Africa World Press).

Chapter 15
Techno: Music and Entrepreneurship in Post-Fordist Detroit

Deborah Che

The music (techno) is just like Detroit, a complete mistake. It's like George Clinton and Kraftwerk stuck in an elevator (Derrick May, quoted in Cosgrove 1988).

While techno pioneer Derrick May's definition of techno as a blend of African-American funk and European electronic music is often repeated, there is a reason why techno music was born in Detroit, and not in Tokyo, Berlin, or London. The development of Detroit techno is tied to the city's deindustrialization (Caux and Torgoff 2000; Lipsitz 2007; Sicko 1999; Williams 2001). Beyond deindustrialization, by examining techno through the lens of Detroit's shift from Fordism to post-Fordism, this chapter finds techno as rooted in Fordist Detroit, but as a quintessential post-Fordist musical production and expression. The chapter will firstly give the historical context for techno through an examination of Detroit's reputation as a center of musical production, consumption, and entrepreneurship and how it was connected to Fordism and Motown and to the city's industrial heritage. Then the chapter will discuss more specifically how techno is tied to Detroit's automotive heritage and its high-wage opportunities under Fordism for blacks and whites and to the city's subsequent deindustrialization. Techno artists drew inspiration from Detroit's mechanized sounds and environment, its past industrial glories, post-industrial problems and future possibilities, which were tied to technological improvements. Through post-Fordist flexible production methods made possible by the falling cost of Japanese electronics and utilization of music industry infrastructure from the Motown past, Detroit's techno innovators created "machine soul" music. The do-it-yourself music production and establishment of record labels, combined with performances in Europe and Japan, fueled the global explosion of techno and the artists' fame. Finally the chapter will look at cultural exhibits and techno music festivals that highlight the link between techno and its Detroit birthplace, which has been largely unrecognized in the USA and in Detroit itself. Techno illustrates that Detroit is a post-industrial, entrepreneurial, creative musical force, and not just one of the industrial, Motown era.

Music in the Motor City: Fordist production and Motown

Detroit has long been the center for numerous entrepreneurial innovations in the transportation and music industries. Henry Ford pioneered one such innovation that changed manufacturing. His moving assembly line for automobile manufacturing assigned each person a specific role, which enabled cars to be produced more efficiently than in the previous artisanal manner. This Fordist production, which spread to other industries, lowered costs with its interchangeable parts, the moving assembly line, strict divisions of labor, and large vertical integrated firms and scale economies (Stutz and Warf 2007). Labor, in exchange for strictly proscribed roles, earned thus unheard of $5 per day in 1914 in Ford's auto plants (Smith 1999). This social contract between capital and labor produced a strong middle class of European immigrants as well as white and African-American migrants from the US South who came to Detroit. These individuals were among the beneficiaries of Fordism as the city was the center of auto production and then the "arsenal of democracy" during World War II. But by the late 1950s, Detroit's downward spiral of deindustrialization, job loss, and depopulation had already begun with the decentralization of auto and other manufacturing jobs from Detroit to its suburbs and to small towns and cities throughout the Great Lakes region and the South. Sugrue (1996) found that all 25 plants the "Big Three" auto producers (Ford, General Motors, and Chrysler) built in the metropolitan Detroit area between 1947 and 1958 were located in the suburbs. Additionally 55 out of 124 manufacturing firms locating in the suburbs from 1950 to 1956 had left Detroit. With Detroit's loss of manufacturing companies and jobs, while being bypassed for new ones, the city's population peaked at just under two million in 1950. The city's population has declined ever since, particularly as whites who had greater mobility to follow jobs out to the suburbs left (Table 15.1). African-Americans were disproportionately affected by decentralization and increased automation as they typically held entry level jobs without seniority (Sugrue 1996).

In this time of shrinking opportunities in manufacturing, Berry Gordy founded Motown Records where music production was built on the Fordist principles he learned on the assembly line at the Ford Wayne Assembly Plant. Gordy explained how he envisioned stars and hit music being produced:

> At the plant cars started out as just a frame, pulled along on conveyor belts until they emerged at the end of the line, brand spanking new cars rolling off the line. I wanted the same concept for my company, only with artists and songs and records. I wanted a place where a kid off the street could walk in one door an unknown and come out another a recording artist – a star (Gordy 1994, 140; quoted in Smith 1999, 14).

Gordy set up the company operations with the strict divisions of labor he observed in the auto industry. Motown's artist assembly took place in adjacent buildings on both sides of Detroit's West Grand Boulevard. The "create" phase involved

Table 15.1   Detroit's population, 1940-2000

Year	Black number	Percentage	White number	Percentage	Total population
1940	149,119	9.2	1,474,333	90.8	1,623,452
1950	300,506	16.2	1,545,847	83.6	1,849,568
1960	482,223	28.9	1,182,970	70.8	1,670,144
1970	659,022	43.6	815,823	54.0	1,511,482
1980	754,274	62.7	402,077	33.4	1,203,339
1990	775,833	75.5	212,804	20.7	1,027,974
2000	770,728	81.0	100,371	10.6	951,270

Source: U.S. Bureau of the Census (1940, 1950, 1960, 1970, 1980, 1990), Characteristics of the Population.http://www.census.gov/prod/www/abs/decennial/index.htm, accessed 1 June 2009; U.S. Bureau of the Census (2000), Census 2000 Summary File 1. http://factfinder.census.gov accessed 1 June 2009.

writing, producing, and recording. Motown songs had interchangeable parts, or rhythm tracks, which would later be incorporated as parts of melodies. The "make" phase consisted of record pressing, inventory management, deliveries, and billing to the distributors. The "sell" phase involved placing records with distributors, getting airplay, marketing, advertising, and promotion, of which the final step on the Hitsville production line was a promotional appearance on the Ed Sullivan show. The Motown production line also included artist development. Training in etiquette, choreography, and stage presence shaped Motown's image and cemented its appeal with Americans of all races as well as international audiences. Another factor in the company's success that further linked the automobile with Motown was that its recording engineers aimed to perfect how a song would sound coming from a car radio, as the company realized how cars fit into youth culture (Smith 1999).

During this Fordist era, Detroit was both an important producer and consumer of music. Given the lack of mass transit, Detroiters spent a lot of time in their cars listening to the radio. Radio stations blared out Motown (the Supremes, Stevie Wonder, Smokey Robinson and the Miracles, Martha and the Vandellas, and the Temptations), as well as also other music which Detroit was famous for, including early punk (Iggy Pop and the Stooges) and rock (MC5, Bob Seger, Ted Nugent). In part due to Fordist manufacturing production in the metropolitan region, Detroit was a great mecca for entertainers. Doug Banker, Ted Nugent's longtime manager, explained,

> If you were a good rock band, you were bigger in Detroit that anywhere else. Detroit is the best concert market in the United States. I've always attributed it to the Big Three, to the auto workers, blue-collar fans who make more money

working in a factory than any other factory job in the United States (McCollum 2003b).

Thus, auto manufacturing jobs in metro Detroit provided a high standard of living with large amounts of disposable income for entertainment.

The legacy of Motown and the Fordist era established the background—the possibilities, opportunities, and challenges for techno, the next major musical innovation to come out of Detroit. Techno music, characterized by the use of analog synthesizers, drum machines, and instrumentals, was the product of middle-class, black beneficiaries of Fordism at the tail end of industrialization in the late 1970s to early 1980s. The college-oriented children and grandchildren of African-Americans who retained jobs in the auto industry, which set the "gold" standard for industrial wages and benefits, fostered the Detroit high school and club party scenes that gave rise to techno. While African-American Detroiters of various income levels were listening to a wide range of music at these parties, middle-class African-American youth in northwest Detroit, the area of the city with the highest household incomes, and in suburbs such as Belleville and Southfield, were prominent among techno's "first wave" of promoters, musicians, and producers. At these parties, an eclectic mix of Parliament-Funkadelic, the Talking Heads, the B-52s, Prince, Donna Summer and Euro-disco acts like Alexander Robotnick was spun. Techno's "first wave" pioneers, the so-called "Belleville Three" who grew up in the western Detroit suburb of Belleville (Figure 15.1) were DJs in the Detroit party scene. The Belleville Three consisted of 1) Juan Atkins, the "Originator," who as part of Cybotron, blended funk and electronic music to generate techno's first major hit, "Alleys of Your Mind"; 2) Derrick May, the "Innovator," who created classic techno tracks such as "Strings of Life" and "Nude Photo"; and 3) Kevin Saunderson, the "Elevator," who popularized techno as part of Inner City which had the major international and crossover hit, "Big Fun" (Sicko 1999). Juan Atkins described the class dimensions of the early African-American party scene as a product of the collective bargaining of the United Auto Workers (UAW) under Fordism:

> My grandfather worked at Ford for 20 years, he was like a career auto worker. A lot of the kids that came up after this integration (of the UAW), they got used to a better way of living. If you had a job at the plant at this time, you were making bucks. And it wasn't like the white guy standing next to you is getting five or ten dollars an hour more than you. Everybody was equal. So what happened is that you've got this environment with kids that come up somewhat snobby, 'cos hey, their parents are making money working at Ford or GM or Chrysler, been elevated to a foreman, maybe even to a white-collar job (Reynolds 1999, 15).

While Fordism's legacy of Motown and auto production provided the historical backdrop to techno created by African-Americans, the shift to post-Fordism and deindustrialization in Detroit was the catalyst for techno production.

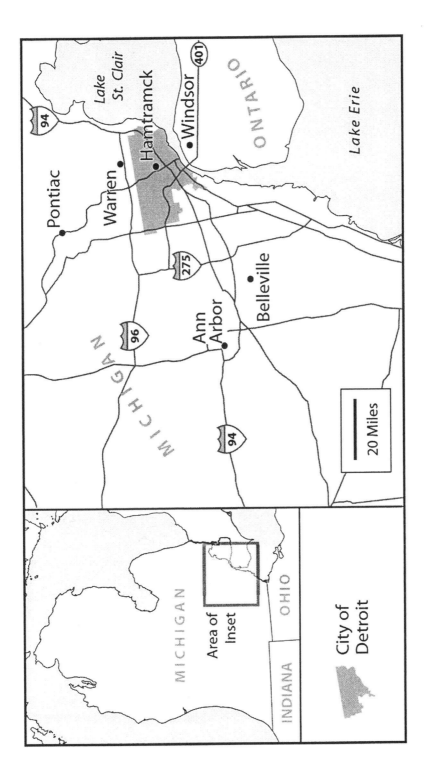

Figure 15.1   Detroit area

From deindustrialization and the perils and possibilities of the machine: Creating techno, a post-Fordist music

The Fordist regime of accumulation, which was supported by complementary social relations such as the Keynesian welfare state, mass production, collective bargaining techniques that ensured high-paid, full employment in households that techno's pioneers had grown up in, and mass consumption fostered by growing disposable incomes, broke down as a result of physical and human limits to production, price-cost squeezes, geoeconomic competition, and the increased mobility of capital (Aglietta 1979; Lipietz 1987). In response to this crisis, capital emphasized flexible technologies, work practices, products and markets. The shift from Fordism to post-Fordism had devastating impacts on Detroit. While periodic downturns occurred by the time Motown left Detroit for Los Angeles in 1972 and when the OPEC oil embargo in 1973 disproportionately impacted the Big Three which produced large vehicles, metro Detroit's formerly cyclical economy appeared to be unidirectional (down) by the late 1970s. The shift to post-Fordist production in the automotive industry involved downsizing, computer aided manufacturing, outsourcing, and offshoring. As a result, overall manufacturing and motor vehicle employment declined 30 percent and 35 percent between 1978 and 1982 in the Detroit MSA. For Detroit's black population, the recession resulted in a shocking 27 percent unemployment rate in 1980. From 1978-1979, Detroit's median household income declined from \$15,623 to \$13,170, while that in the USA as a whole rose ten percent (Ross and Trachte 1985). The crisis in capitalism and auto manufacturing in Detroit provided the impetus for techno music, which was produced in response to the conditions in Detroit.

According to Sheridan (2001), techno began as an expression of black Detroiters' experience with the effects of globalization that was simultaneously realism (an expression of what was going on) and escapism (the hope for a better future through technology). Detroit techno has been described as hopeful, euphoric, and simultaneously anxious in its finding of "beauty in decay" in the deindustrialized city (Figure 15.2). Derrick May attributed the dreaminess of Detroit techno to the abandoned buildings, or "… the emptiness in the city that puts the wholeness in the music" (Reynolds 1999, 21). But this was not a purely romantic reflection, as techno conveyed the post-industrial, post-Fordist realities in Detroit. The music mirrored the real aftermath of the aforementioned job losses and declining household incomes. As May explained,

> This city is in total devastation. It is going through the biggest change in its history. Detroit is passing through its third wave, a social dynamic which nobody outside this city can understand. Factories are closing, people are drifting away, and kids are killing each other for fun. The whole order has broken down. If our music is a soundtrack to all that, I hope it makes people understand what kind of disintegration we're dealing with (Cosgrove 1988, 89).

While reflecting the bleakness of post-industrial Detroit through synthesized sounds, techno's pioneering artists believed that technology and machines, which had led to manufacturing job losses since the 1950s, would also lead to a better future for Detroit. As a student at Belleville High School, techno's "Originator," Juan Atkins learned the theories of Alvin Toffler in a future studies class. Toffler in his book, *The Third Wave*, wrote about "techno rebels" who recognize and use technology that is sophisticated without necessarily being big, costly, or complex (Sicko 1999). Atkins, who picked up the term techno and used it to coin the music he would develop, saw the perils and possibilities of technology since, "When the new technology came in, Detroit collapsed as an industrial city, but Detroit is techno city. It's getting better, it's coming back around" (Savage 1993, 20; quoted in Sheridan 2001, 148). Perhaps the other-worldly looking Renaissance Center featured on several techno record covers epitomized techno city, a future Detroit landscape that contrasted with the remnants of the city's former industrial glories like the Packard Plant (Figure 15.2). A historian of techno, Sicko (1999) considered the common thread connecting Toffler and Detroit techno musicians to be their belief in the power of the individual and personal visions of utopia which may be attained through the use of such technology. Techno hearkened back to earlier traditions in African American music which linked the future, transformation, liberation, technology, and pan-Africanism (for example, Sun-Ra's Intergalactic Research Arkestra, George Clinton's P-Funk Mothership Connection) (Connell and Gibson 2003; Williams 2001). In contrast with British punk rock, techno, which emphasized the machine as part of Detroit's past, present, and future, thus had a more optimistic view of the post-industrial future.

The sound of machines, part of the environment in Detroit, was also key to techno music. Carl Craig, who was part of the "second wave" of Detroit techno artists following the Belleville Three, highlighted the connection between industrial Detroit and techno in noting, "Our backgrounds of having relatives, friends, parents that work in factories, from the Motor City where the machine is king gave fascination with electronics like synthesizers, drum machines" (Detroit Historical Museum 2003). While Detroit's techno artists were not working in factories, which were employing an ever-declining number of people, Detroit's mechanized aural and visual environment was an integral part of techno, as Derrick May explained:

> We were led to create this music unconsciously. We got the idea of machines and we made our own sounds. All the sounds came from the world of machinery, industry, electronics, our environment. In this busted-down town, techno was an attempt to move in reverse and become the soundtrack for an imaginary world where people would benefit from the machine instead of being alienated by it. A world where technology would no longer trample on creative and cultural activity (Caux and Torgoff 2000).

Figure 15.2a Future: Renaissance Center

Hence the machine could also benefit people, not only replace them in the production process.

In addition to machines and the landscape reflecting the flight of capital, jobs, and population, Detroit radio played a major role in the creation of techno. Local radio station/programs receptive to new and/or local trends are commonly

Figure 15.2b Past: Packard Plant

important in the formation of music "scenes." Many of techno's early artists have cited DJ Charles Johnson, better known as the Electrifying Mojo, as a major influence. The Electrifying Mojo's free-form radio programs, playing black and white artists such as the B-52's, Talking Heads, Funkadelic, Prince, and the J. Geils Band, also looked to a better future where race did not matter. This outlook was truly optimistic in a city that experienced a major race riot in 1943, physical and verbal attacks on blacks moving into all-white neighborhoods in the 1950s and 1960s, followed by rapid white flight to the suburbs (Table 15.1) (Sugrue 1996), and the 1967 riot/rebellion,[1] which led to 43 deaths and $40-80 million in damage (*The Detroit Free Press* 2007), accelerated white flight (Table 15.1), and negatively impacted Detroit's image. Starting on Ann Arbor's rock radio station

---

1   The disturbances in Detroit from 24-30 July 1967 have alternatively been termed a riot or rebellion. They began with a police raid on a blind pig, or unlicensed bar, in an African-American neighborhood. The raid escalated into looting, shooting, and the burning of large sections of Detroit. The national and local media and government officials referred to the events as a riot as did most whites. Many white Detroiters give the riot as their reason for leaving Detroit. However in the African-American community, the term rebellion (against police brutality, inequality in housing and employment) is frequently used to denote the same disturbances (Lee 2007).

WAAM in the early 1970s, the Electrifying Mojo expressed a vision of radio and life on the airwaves that continued on a half-dozen radio stations in Detroit from the 1970s-1990s, which influenced listeners throughout metropolitan Detroit and Windsor, Ontario who were to become techno artists (Figure 15.1):

> I was in Ann Arbor standing on the corner of Stadium and University ... I was thinking about what the mission of radio should be. I saw all of these different cultures, ethnicities passing by me ... Old people, young people, black people, white people, Native Americans – people from the whole world. I was thinking about how radio stations fight for market share. They look at radio through this narrow prism ... During the daytime the wall of separation was so pervasive. It makes you act as who you are not. Then I noticed black radio stations, white radio stations, it was like apartheid on the dial. They have this niche audience in mind and they put blinders on and they're not going to see above these blinders ... So, with radio and me, I'm taking all of these disjointed, disconnected pieces. I'm taking all these cultures and I'm going to smash them together in a blender and see what happens ... I took Elton John, the B-52s, and put them with Prince, George Clinton and the Gap Band. Stevie Nicks alongside Aretha Franklin. All music is interrelated and the simple thing is to find the point where it interfaces and I think that's where people miss the point ... I took the music from everybody and put it all together. It came to a point where the white people say I could get into 'One Nation Under a Groove' (by Parliament-Funkadelic) and the black people said they could get into 'Mesopotamia' (by the B-52s) (Patricola 2005a, 45-6).

On his show, the Midnight Funk Association, the Electrifying Mojo "smashed" music together, which exposed techno's pioneers to different musical genres. For instance, Mojo regularly played and popularized Kraftwerk, the electronic music group from Düsseldorf, Germany, another automobile manufacturing city, whose members were fans of Motown and James Brown and aimed to create music combining American rhythm and European harmony and melody (Sicko 1999). Kraftwerk appealed to African-American Detroiters because of their clean, precise sounds that were "so stiff, they were funky," while to Derrick May they sounded like "somebody making music with hammers and nails" or "like a tribal calling" to people in the industrial city (Reynolds 1999, 13-14). May, in reflecting upon radio in this era, said,

> Detroit was much further ahead of most black communities outside of New York. We had Kraftwerk, Jimi Hendrix, and Prince on the radio all day long, one song after another. There were so many interesting people coming out of nowhere, white and black. The music, in one sense, was romantic, emotional and soulful. In another sense, it was colorless. It was beautiful (Darling 2001).

Techno was also born from this beautiful mix of black and white music on the Electrifying Mojo's show. In overcoming separation on the airwaves and on the ground in the nation's most segregated metro area, Mojo placed George Clinton and Kraftwerk together in the creative minds of young Detroiters, who then fused them to create techno. In addition to Mojo, Jeff Mills, whose skills on multiple turntables earned him the moniker of The Wizard, likewise influenced radio listeners to become techno artists themselves. Through the influence of local radio, African-Americans produced techno. This transcultural musical form in which its creators were influenced by local and global cultural traditions and economic forces (Kong 1996) also exemplified Gilroy's (1993) Black Atlantic concept in which compound black cultures have been formed through hybridity and the intermixture of ideas from disparate sources on both sides of the Atlantic.

The flexible, do-it-yourself production of techno

As declining opportunities in manufacturing made it increasingly unlikely that they would follow in the footsteps of their grandparents and parents, Detroit's techno artists created their own music and recording labels, utilizing flexible production methods. The falling cost of technology and musical equipment, specifically Japanese sequencers and synthesizers, enabled a do-it-yourself, experimental approach to creating techno music in home studios (Cosgrove 1988). "Second wave" techno artist Richie Hawtin from Windsor, Ontario explained how important technological innovations were in his music production:

> The biggest innovation that enables me to be here was the accessibility of equipment and the idea of a home studio. Before techno and electronic music came along, if you wanted to record an album you had to go to a $100,000 studio, you had to have producers, engineers, and all that shit. Part of the explosion of electronic music was the idea that one or two people can be in their bedrooms with enough instruments at an affordable price creating their own albums (Osselaer 2001).

Many of Detroit's classic, early techno hits, were produced in home recording studios, which contrasted greatly with Motown's Fordist assembly line. With Japanese electronics costing only $1,500-2,000, the financial barriers to entry into cottage-industry style techno music production were extremely low (Graff 1993). Unconstrained by the limitations of studio costs, Detroit's techno rebels could create music with this low-level technology. The machines freed the techno artists from uncertain futures in light of the recession and crisis of early 1980s Detroit. Mike Banks, co-founder of the influential and political techno production company Underground Resistance, noted as much, saying "... maybe there's some technology there that could help you escape whatever messed-up situation you're in. For us, it was Japanese electronics. You know it was just simple. It brought it out of the big studio into the bedroom" (Copeland 2004).

Unlike vertically integrated firms such as Ford and Motown which controlled and/or operated all parts of the production process, the mastering and manufacturing of techno music produced in home studios were handled by an infrastructure of music industry experts dating back to the Motown era who had their own companies separate from those of the techno artists themselves. For instance, Ron Murphy, a sound engineer who worked with Motown and has cut most of Detroit techno artists' masters as his expertise in bringing up the bass rather than cutting "as is," made the music even hotter on the turntables. Murphy's innovation in cutting records mirrored that of the artists he worked with. For Jeff Mills's "Rings of Saturn" release, he cut the vinyl, which had giant spaces on it to resemble the planet, so the record played each song in separate loop grooves instead of from the outside in (Patricola 2005b). In addition to Murphy, Archer Records, the only vinyl-pressing plant in Detroit, also has a mutually beneficial relationship with techno. Archer has been as busy doing batch 12" production for techno artists as it was during the 1970s (Warshaw 2001).

Like Berry Gordy decades earlier, the techno artists were entrepreneurs, establishing labels to distribute their music. At one time a stretch of Gratiot Avenue in Detroit's Eastern Market neighborhood was known as "Techno Boulevard" because the Belleville Three had located the headquarters of their respective labels, Metroplex (Juan Atkins), Transmat (Derrick May) and KMS (Kevin Saunderson) there (Sicko 1999). Most of these records were sold in Europe, where techno took off with sales of the aforementioned Inner City hit, "Big Fun" and a Virgin UK compilation, *Techno! The New Dance Sound of Detroit.* The 100,000 records a year Juan Atkins sold under his Metroplex Records label were fueled by overseas performances (*The Detroit News* 1999). Those overseas performances opened up a new world to the techno artists as they could experience their global impact on electronic music. Although the Belleville Three grew up in a predominantly white suburb, living and working in Detroit highlighted the realities of segregation. Juan Atkins mused on the experience, "If you're a kid in Detroit, you might never even have to see a white person, unless they're on TV. The first time I went to the U.K., man, I played for five thousand white kids" (Reynolds 1999, 71). Traveling multiple times a month to Europe and Asia for club performances starting at $3,000 a gig (McCollum 1998) became the norm for the Belleville Three, while they remained obscure in the USA except for a small, knowing techno crowd. The traveling schedules of the Belleville Three also made it difficult for "second wave" artists to get heard and possibly signed to their labels. As a result, artists such as Carl Craig and Richie Hawtin started their own labels (Planet E and Plus 8 respectively). Carl Craig started Planet E in 1991 with an initial $1,500 investment, but came to generate $500,000 annually, mainly through UK and German record sales. While Detroit's auto brands have faltered, the Detroit brand became a strong selling point in techno as Carl Craig noted, "People in America don't seem to be interested in techno, but overseas, they eat it up. They'll buy anything that comes from a Detroit artist, because Detroit is widely acknowledged as the birthplace of techno." Mike Himes, owner of Record Time Distributors in Roseville, distributed records for 40 local techno labels with about 80 percent of his business coming from overseas. In describing the post Fordist, do-it-yourself approach of Detroit techno artists he said, "A bunch of these kids have made a lot of money in Europe. In most cases, the artists are also the owners of their record companies. I don't think they had any intention of becoming record executives, but that's how things have turned out" (*The Detroit News* 1999). The artists, from the Belleville Three to the "second wave," "third wave," and beyond, created both a music genre and an industry in a vacuum left by the departure of Motown and automobile industry jobs.

In addition to Record Time and the record labels, Submerge, a company started in 1992 by Mike Banks, co-founder of Underground Resistance, and Christa Robinson, handled invoicing, manufacturing, and distribution for many of the labels owned by jet-setting DJs (Detroit Historical Museum 2003). Submerge exported about 80 percent of the city's electronic music labels to retail customers abroad (Casper 2004). Just as in its recording and record manufacturing, techno's distribution and promotion followed a very different model than Motown, which tightly controlled every part of the production process, including its artists. Through Submerge's collective "umbrella" environment artists could retain artistic independence and survive without major labels (Sheridan 2001) as Submerge handled much of the day-to-day business functions. Detroit techno, at this early stage in the product life cycle, grew rapidly as its artists developed and commercialized this musical expression through their own labels and the Submerge collective.

Recognizing techno in the motherland

While techno exploded in Europe and Japan, which were more open to dance music than the rock and later rap-oriented USA, Detroit techno artists were relatively unknown stateside. By performing and selling records for years in Europe, they expanded the fan base, influenced future techno artists there, and helped birth what Connell and Gibson (2003) call a transnational soundscape. In places influenced by Detroit such as the UK, Germany, the Netherlands, and Japan, techno has taken on different forms. However, the Detroit techno artists left domestic market undeveloped. As a result, when techno hit the USA as part of the rave scene, it was the latter-day European artists such as the Chemical Brothers and Fatboy Slim and Euro-Americans such as Moby that achieved notoriety. Many Americans thus believed that techno, the music created in Detroit as the fusion of man (soul) and machine (electronics) that reflected the city's post-industrial realities, originated in Europe. While it is well known in Europe and Japan that techno came from Detroit and is a "black art form ... among the young white demographic that is into electronic music in the U.S. [it is not]" (McCollum 1998).

In part to rectify this knowledge gap and to recognize techno in its Detroit birthplace, electronic music festivals and cultural events recently have taken place

Figure 15.3   Carl Craig

in the city. On Memorial Day weekend 2000, Hart Plaza in downtown Detroit was the site of the first Detroit Electronic Music Festival (DEMF). On the eve of this first festival, Derrick May hoped it "could open the eyes of a lot of black folks in Detroit who don't realize this music has roots here" (McCollum 2000). His hopes were realized when this event attracted 1.5 million people, both from around the world who traveled to the "motherland" or "mecca" of electronic music and from Detroit itself (Case 2000; McCollum et al. 2001). The 2000 festival and subsequent ones every Memorial Day weekend since have given exposure to Detroit's musicians (Figure 15.3). Carl Craig, the 2000 festival producer and co-organizer, commented on how the festivals finally brought the music to a huge home audience:

> When I was standing up on stage and Terrence (Parker) and Stacey Pullen playing, and as it got nighttime, all the people in the bowl, it just became this feeling, it was all spirit. The last day, standing up there with all the artists, we're all looking and we're all choked up and we're all just, 'Wow.' We never believed this would ever happen. We'd been traveling all over the world, never saw the recognition here and then this is what it is, firsthand (Giannini 2001).

Techno has additionally been recognized in its birthplace in a unique exhibit that ran from 17 January 2003 to 15 August 2004 at the Detroit Historical Museum entitled, "Techno: Detroit's Gift to the World." In welcoming visitors, this exhibit situated techno as part of local history and Detroit's entrepreneurial vision:

> Ever since its founding in 1701 Detroit has been a breeding ground for people that have gone against the norm, set trends, and changed the world with a belief in the entrepreneurial spirit. Welcome to Techno: Detroit's Gift to the World as we create "LIVING HISTORY" by documenting the history of techno music and its origins in Detroit. The story begins in the early 1980s with Juan Atkins who not only created a new style of music, but also had the vision and belief in this style to elevate it to international heights. The story continues as we follow many other Detroit artists around the world and return to Detroit to celebrate their achievements in creating a global phenomenon (Detroit Historical Museum 2003).

The exhibit detailed the twenty year history of techno from its underground roots, its popularity first in Europe and then globally, and finally its recognition in Detroit with the first DEMF. It told the stories of the DJs, producers, and other key figures in the manufacturing and promotion of techno music by exhibiting equipment, records, fliers, photos, and video segments with artist interviews. For later museum-goers, the exhibit included interactive displays that replicated an underground dance club and makeshift studio which allowed visitors to mix their own techno track (*The Detroit Historical Society Newsletter* 2003; Case 2003).

While the Detroit Historical Museum had extensive permanent exhibits on Detroit's settlement and development, including its cobblestone re-creation of the pre-automobile streets of Old Detroit and an assembly line auto drop, the techno exhibit augments recent history offerings. Sulaiman Mausi, a Detroit native and the exhibit's developer, noted that such recent history was just as relevant, saying, "… we don't have to keep talking about Henry Ford or … stuff that happened 100 years ago. History was created yesterday, last week, last year" (Graff 2003). To persuade museum board members to approve an exhibit which might be considered unconventional or "edgy," Mausi stressed how techno's international appeal and impact on electronic music was akin to Motown's influence on 1960s and early 1970s pop and rock music. Bob Bury, executive director of the Detroit Historical Society, felt the exhibit could help the Detroit Historical Museum attract new visitors, while being "entertaining, educational and provocative" in its telling the important story of how Detroit changed the world through music. For the museum, the exhibit was a way to be relevant to today's Detroit population and to "get in the game and be seen as a place people look to for a cultural experience." According to Dan Sicko, the Detroit Historical Museum exhibit also "legitimizes its (techno's) role in Detroit cultural history" (Provenzano 2003). Likewise, Kevin Saunderson noted the exhibit is "more than appreciation, really – it makes it history. It's not just people writing or talking about it. It's real" (McCollum 2003a).

By telling the history of techno, a music created to reflect, interpret and move beyond Detroit's "Great Depression" of the 1980s, the hope was to inspire today's youth. Exhibit coordinator Sulaiman Mausi wanted young viewers to see what their peers did when they faced a seemingly hopeless situation twenty years ago. Mausi explained,

> We want kids to say, 'If he can do it, I can do it.' Another thing that's important to capture is their (techno artists') entrepreneurial spirit. When Juan (Atkins) couldn't find anyone to put out his records, he did it himself. It's about the music, but it's also about the accomplishment of creating something that was not there (Gorell 2003).

When proposing the exhibit, Mausi thought techno would be the right vehicle to connect to young Detroiters as "… they can see, 'Wow, this guy went to my high school'" (Graff 2003). For this reason, each DJ/producer's high school is included in the displays. Inspiring youth, Detroit's future, is important to the techno artists who were themselves inspired by the Electrifying Mojo's and the Wizard's radio shows, to fashion their own music, record labels, and futures. Today's youth of Detroit have little opportunity to hear techno (or any music than the most mainstream) given the consolidation and corporatization of radio. They may not be familiar with or like techno, whose black roots are not widely recognized in the USA. But the exhibit's aim is to inspire youth to believe that they too can be creative and successful, whether or not that creativity is expressed through music or other avenues. Alan Oldham (DJ T-1000) hopefully says,

> A lot of young kids come through here, a lot of field trips from schools. And the hope is that a younger person will go into the exhibit and see all these young, black men who've accomplished this, and maybe it'll spark someone else not necessarily to be a musician, but to say, 'If these guys did this, I can do what I wanna do' (Rayner 2003).

The lessons of techno

Perhaps like at no other point since the early 1980s, today's (2009) metropolitan Detroit and Michigan faces major economic challenges. The city of Detroit, which has continued to lose jobs and population, now has less than one million residents. In 2000, 81 percent of its 951,270 residents were black, while only 10.6 percent were white (Table 15.1). In contrast, the seven counties surrounding the city, with most of the metro area's jobs and population, are 87 percent white (James 2004). While Detroit has lost most of its auto and other manufacturing jobs, the restructuring of the US auto industry is now heavily impacting its suburbs as well as other parts of the so-called Rust Belt. Through buyouts and plant closures, Ford, General Motors, and Chrysler are trying to bring their labor force and total production in line with declining sales. While Michigan is less dependent on the auto industry than in the early 1980s, compared to other states it is still heavily reliant on manufacturing. Hence it is mired in a one-state recession, decoupled from much of the US economy. Since 2001, Michigan has lost 305,000 jobs, with an estimated 40 percent of those from automakers and their suppliers (ElBoghdady 2007). According to "A Region in Turbulence and Transition," a recent report issued by the Southeast Michigan Council of Governments, dramatic losses in the domestic auto industry will keep metropolitan Detroit in an economic crisis for at least a decade. The current problems are structural, not cyclical like past ones in the 1970s, 1980s and 1990s, when the Big Three pulled the region out of recessions. Given the Big Three's rising costs and declining market share, only the "rise of a more educated, skilled work force" and a "new Michigan," can alter the report's chilling predictions (Wilkinson 2007).

Thus, the parable of techno is particularly valuable today as Detroit and Michigan look for a way forward. Techno showed that in the midst of despair, creativity can thrive. Powered by an entrepreneurial spirit, African-Americans produced a multi-million dollar industry. While machines and technology made certain manufacturing jobs redundant, techno exemplified the importance of embracing technology and new ideas for a post-industrial future. In discussing the analog-digital debate in techno, Mike Banks of Underground Resistance spoke of the power of machines and technology:

> If you really are making futuristic music, then you gotta let the future in your heart too man. Back in the pre- MIDI[2] days I remember arguing with musicians about MIDI and the advancement of MIDI and I imagine people with their horse and buggy had the same argument. You know when you see some technology that's obviously forward, it's good to embrace it man. For us, it's kind of like how the cotton gin was for slavery. It's like that technology really freed people. You know it's easier to work a machine than it was to work 300 mugs. You know, the machine is cheaper man. So economically the machine really helped us free us…so I know sometimes guys get off to the analog digital debate, but don't be a dinosaur - that's my thing (Copeland 2004).

Banks's words go beyond the debate in techno to thinking about the region's need to change and adapt to new technologies. However, technology and the machine are only means to an end. At its roots, techno is about creativity, spirit, independence, resistance, and persistence, which were expressed through machines. While techno or the music industry in Detroit cannot be expected to replace thousands of auto industry jobs, metro Detroit can no longer place its future in the hands of one industry. Future changes that will involve "multiple technos" or new enterprises

---

2   MIDI, or Musical Instrument Digital Interface, is an industry-standard electronic communications protocol that enables electronic musical instruments, computers, and other equipment to communicate, control and synchronize with each other in real time.

are necessary. Given Detroit's long tradition of being a creative music force stretching from Motown, early punk, Detroit rock city to techno, latter-day garage rock, hip-hop, and rap, some of these enterprises will likely involve music. Perhaps the Electrifying Mojo's signature sign-off during his nightly words of wisdom segment best captures the spirit of Detroit: "When you feel like you're nearing the end of your rope, don't slide off. Tie a knot. Keep hanging, and remember ... that ain't nobody bad like you" (Patricola 2005a).

References

Aglietta, M. (1979), *A Theory of Capitalist Regulation* (London: New Left Books).

Case, W. (2000), "Electronic Music Fest Draws 1.5 Million Fans," *The Detroit News* 31 May, p. 1.

Case, W. (2003), "Museum Tells Techno's Tale: Historical Venue Gives Homegrown Music Overdue Attention," *The Detroit News* 16 January, p. 1.

Casper, K. (2004), "In the Flesh Again: Mike Banks and the Elusive Underground Resistance Set for First Live Detroit Gig Since 1992," (published online 26 May 2004) <http://www.metrotimes.com/editorial/story.asp?id=6275>, accessed 28 June 2007.

Caux, J. and Torgoff, L.-S. (2000), "Detroit: The Agony and the Ecstasy: Motown's Techno Unconscious," *Art Press* 260, 34-7.

Connell, J. and Gibson, C. (2003), *Sound Tracks: Popular Music, Identity, and Place* (London: Routledge).

Copeland, L. (2004), Interview with Mad Mike Banks on the Liz Copeland Program, WDET-FM 101.9, 8 Nov.

Cosgrove, S. (1988), "Seventh City Techno," *The Face* 97, 86-9.

Darling, C. (2001), "Black Youths Gave Quiet Birth to Techno," *The Miami Herald*, 15 April, p. 1M.

Detroit Free Press, The (2007), "The Riot by the Numbers," (published online 20 July 2007) <http://www.freep.com/apps/pbcs.dll/article?AID=/20070720/NEWS01/707200389/0/NEWS01&theme=DETROITRIOT072007>, accessed 23 July 2007.

Detroit Historical Museum. (2003), Techno: Detroit's Gift to the World, museum exhibit, 17 January, 2003 – 15 August, 2004.

Detroit Historical Society Newsletter, The (2003), "Techno Exhibit Draws Big Crowds, Worldwide Attention," 10:1, 1, 11.

Detroit News, The (1999), "Motown's Music Rules Again: But Detroit's Techno Beat Finds Fans Mainly in Europe," 25 February, p. 1.

Detroit News, The (2007), Interactive: Population Change in Metro Detroit. (published online 19 July 2007) <http://detnews.com/apps/pbcs.dll/article?AID=/20070719/METRO/70718001>, accessed 24 July 2007.

ElBoghdady, D. (2007), "Housing Crisis Knocks Loudly in Michigan: Foreclosures Hit Record Numbers as Region Continues to Lose Jobs," (published online 31 March 2007) <www.washingtonpost.com/wp-dyn/content/article/2007/03/30/AR2007033002127.html>, accessed 2 April 2007.

Giannini, M. (2001), "Electric Heaven: Could It Be, Would It Be, the Festival of Our Dreams?" (published online 22 May 2001) <http://www.metrotimes.com/editorial/story.asp?id=1825>, accessed 20 May 2004.

Gilroy, P. (1993), *The Black Atlantic: Modernity and Double Consciousness* (Cambridge, MA: Harvard University Press).

Gordy, B. (1994), *To Be Loved: The Music, the Magic, the Memories of Motown: An Autobiography* (New York: Warner Books).

Gorell, R. (2003), "Permanent Record: Jeff Mills Talks Detroit Techno and the Exhibit That Hopes to Explain It; For Once, the Music's History Will Be Open to the Public for Debate," (published online 15 January 2003) <http://www.metrotimes.com/editorial/story.asp?id=4496>, accessed 20 May 2004.

Graff, G. (1993), "Sonic Boom at Clubs and Rave Parties, Techno Is the Hard-Driving Dance Music of Choice," *Detroit Free Press* 5 November, p. 1E.

—— (2003), "Birth of a Techno Nation: Exhibit Traces the History of a Music Revolution That Started in Motown and Spread Across the Globe," *The Oakland Press*, 17 January 2003.

Hines, A.H. et al. (2001), *TechniColor: Race, Technology, and Everyday Life* (New York: NYU Press).

James, S. (2004), "A Frenzy of Change: How Northland, Now 50, Jump-Started Suburbs' Growth," (published online 18 March 2004) <http://www.freep.com/news/metro/burbs18_20040318.htm>, accessed 9 August 2004.

Kong, L. (1996), "Popular Music in Singapore: Exploring Local Cultures, Global Resources, and Regional Identities," *Environment and Planning D: Society and Space* 14:3, 273-92.

Lee, A. (2007), "Riot or Rebellion? Detroiters Don't Agree; Whites Call 'Riots' Focal Point of Detroit's Decline, but Blacks See 'Rebellion' Against Rooted Inequities," (published online 19 July 2007) <http://www.detnews.com/apps/pbcs.dll/article?AID=/20070719/METRO/707190418/-1/ARCHIVE>, accessed 21 July 2007.

Lipietz, A. (1987), *Mirages and Miracles: The Crises of Global Fordism* (London: Verso).

Lipsitz, G. (2007), *Footsteps in the Dark: The Hidden Histories of Popular Music* (Minneapolis: University of Minnesota Press).

McCollum, B. (1998), "U.S. Pulses to Detroit Techno; 3 Pioneering Artists Already Known Throughout Europe for Futuristic Music That Mixes Funk, Synthesized Sounds," *Detroit Free Press*, 16 September, p. 1A.

—— (2000), "Higher Volume Local Musician Brings Electronic Music Festival to Detroit and, He Hopes, into the Mainstream," *Detroit Free Press*, 1 May, p. 1A.

—— (2003a), "Techno Goes Legit; Multimedia Project Gives Music Stamp of Approval," *Detroit Free Press*, 12 January, p. 1F.

—— (2003b), "Rock City: Bands Love to Play Metro Detroit Venues, and Fans Love to Party With Them," *Detroit Free Press*, 10 August, p. 1E.

McCollum, B. et al. (2001), "Techno Know-How," *Detroit Free Press*, 20 May, p. 4G.

Osselaer, J. (2001), "Richie Hawtin (August 2001 Interview)," (published online 10 August 2001) <http://technotourist.org/modules.php?op=modload&name=Sections&file=index&req=viewarticle&artid=4>, accessed 19 May 2004.

Patricola, V. (2005a), "The Electrifying Mojo: Part I - The Mission," *Detroit Electonic Quarterly* 3, 44-51.

—— (2005b), "Ron Murphy: Still the Master," *Detroit Electronic Quarterly* 1, 24-6.

Provenzano, F. (2003), "Techno Goes Legit: Museum Hopes Exhibit Updates Its Stodgy Image," *Knight Ridder/Tribune News Service* 15 January, K2143.

Rayner, B. (2003), "Tracking Techno in Motor City," *Toronto Star* 26 January, p. D3.

Reynolds, S. (1999), *Generation Ecstasy: into the World of Techno and Rave Culture* (New York: Routledge).

Ross, R. and Trachte, K. (1985), "The Crisis of Detroit and the Emergence of Global Capitalism," *International Journal of Urban and Regional Research* 9:2, 186-217.

Savage, J. (1993), "Machine Soul: A History of Techno," *The Village Voice Summer 1993 Rock and Roll Quarterly*, 20 July, 18-21.

Sheridan, D.M. (2001), *Narrative and Counter-narrative in Detroit*. Ph.D. Dissertation (East Lansing: Michigan State University).

Sicko, D. (1999), *Techno Rebels: The Renegades of Electronic Funk* (New York: Billboard Books).

Smith, S.E. (1999), *Dancing in the Street: Motown and the Cultural Politics of Detroit* (Cambridge, MA: Harvard University Press).

Stutz, F.P. and Warf, B. (2007), *The World Economy: Resources, Location, Trade and Development* Fifth Edition (Upper Saddle River, NJ: Pearson Prentice Hall).

Sugrue, T.J. (1996), *The Origins of the Urban Crisis: Race and Inequality in Postwar Detroit* (Princeton: University Press).

Warshaw, A. (2001), "Keepin' Vinyl Alive: Cutting Records the Good Old Way for the Cognoscenti," (published online 27 February 2001) <http://www.metrotimes.com/editorial/story.asp?id=1376>, accessed 30 June 2007.

Wilkinson, M. (2007), "Michigan's Pain Far From Over," (published online 30 March2007)<http://www.detnews.com/apps/pbcs.dll/article?AID=/20070330/METRO/703300437>, accessed 30 March 2007.

Williams, B. (2001), "Black Secret Technology: Detroit Techno and the Information Age," in Hines et al. (eds.).

Chapter 16
The Production of Contemporary Christian Music: A Geographical Perspective

John Lindenbaum

Contemporary Christian Music, or CCM, is one of the fastest growing segments of the music industry at a time when the record industry is in a state of crisis with declining sales (Hiatt and Serpick 2007). It is also a set of evangelical Christian cultural practices operating in a moment rife with rhetoric of cultural and political polarization. On the *Billboard* charts, at huge summer festivals, in church services, in movies, and on radio stations throughout the United States, the combination of Christian lyrics and musical styles from hip hop to adult contemporary contribute to a particular music geography. Though relatively neglected in geography and academic literature on the whole, CCM can elucidate much about the music industry, evangelical Christianity in the US, and the convergence of religion and commerce.

What follows is an analysis of CCM from two perspectives. Firstly, the geography of CCM is produced by the social practices of the musicians, recording engineers, record company executives, radio programmers, booking agents, and journalists responsible for the creation of the music. The second aspect of CCM entails the definition of the genre. CCM does not exist as a natural, discrete category; rather, it is created through a series of delineating acts. The process of defining what qualifies as CCM—which artists, which lyrics, which conceptions of Christianity—is also part of its production, and contributes to the geography of evangelical Christianity in the United States. Considering CCM to be both materially and symbolically produced helps elucidate the genre.

As CCM is an understudied musical category, the objective of this chapter is to provide a broad overview of the geography of CCM, including the definition of the CCM niche marketing category, the evangelical Christian theology expressed in its lyrics, the advent of the CCM industry, the corporatization of its record companies and consequent prioritization of business over ministerial goals, CCM's industrial centralization in Nashville, the corporate consolidation and geographic expansion of CCM radio, and the problematic binary sacred/secular tropes used to describe CCM. Related aspects of CCM are not considered here, including how people listen to recorded music, the experience of attending a live concert, the meanings it has for listeners, the formation of religious communities. These topics will be addressed in future research.

What is CCM?

Unlike genres of music such as blues, country, or rock, CCM does not adhere to one particular stylistic palate. CCM includes many styles of music, such as the adult contemporary of Mercy Me, the hip hop inflected pop of TobyMac, and the screamo (emotional hardcore punk with screaming) of Underoath. Which artists and songs qualify as CCM is a matter of some debate. Possible defining factors include the known beliefs of the artist (whether he or she is a practicing Christian), the lyrical content of the music, and the Christian identity of a record company, known colloquially as a "label."

However, I would argue that the most apt definition of CCM is that it includes popular styles of music aimed at the so-called Christian market. This Christian market, a hypothetical group of consumers, is represented by gatekeepers such as Christian radio stations and Christian retailers. CCM, thus, is a marketing term more than a referent to a particular sound or lyrical topic. For example, if country stars Alan Jackson, Alabama, or Carrie Underwood market a song to retail outlets such as Christian bookstores and Christian radio stations, that country song becomes CCM. Similarly, the hip hop-influenced hard rock of general market chart-toppers P.O.D. can also be heard on Christian radio stations. However, U2, a band of Christians singing at-times biblical lyrics, are not considered to be CCM since they eschew association with the Christian market. U2 does not appear on the Christian charts or on Christian radio, but a 2004 cover of U2's "Pride (in the Name of Love)" by the Christian-marketed band Delirious? did.

The term CCM generally does not refer to traditional "black" gospel such as Fred Hammond, nor does it encompass "white" Southern gospel such as Bill Gaither. Gospel musical styles, along with children's albums and other religious recordings, join CCM under the larger category of Christian/Gospel. The distinction between CCM and traditional or Southern gospel is relatively easy to make, since the latter two categories are recognizable *styles* of music. However, the difference between CCM and general market popular music is much murkier, since the same musical styles appear in both categories.

A host of players in the CCM industry determine what qualifies as CCM. The Gospel Music Association, which represents the CCM music industry as well as the Southern and traditional gospel industry, administers its Grammy-like Dove awards to music with 50 percent or more identifiably Christian lyrical content. Music industry trade journal *Billboard's* charts for Christian music require 50 percent Christian content, significant national Christian radio airplay, or 25 percent of a record's first week sales to be in Christian retail outlets (Styll 2007). Record companies decide whether an artist should be aimed at the Christian market. Christian bookstores, which were until recently the primary retail outlet for CCM, decide which albums to carry. Christian radio stations or their parent companies choose which artists to play, based on content, popularity and whether record companies send a new release to them. The organizers of Christian summer music festivals decide which artists can perform. Magazines, such as *CCM Magazine*, cover some acts but ignore others. Lastly, artists declare themselves to be "in" or "out" of the Christian market. There are many fringe cases, including the proverbial "artists who are Christian but not Christian artists." As for any socially produced categorization, the boundaries of CCM are blurry and in flux.

While there is no single CCM sound or style, many of the lyrics and on-stage messages of CCM do reflect an evangelical Christian worldview rather than a Catholic, "mainline" Protestant, or generic Christian perspective. This theology stresses a personal relationship with Jesus, proselytizing, Biblical inerrancy, and being born again. Many CCM songs, particularly the radio-friendly adult contemporary and pop variants, express this personal theology in so-called "Jesus is my girlfriend" songs, which address God in the same way a Top 40 song would address a new crush. The "Jesus count" or "Jesus-per-minute" of a song's lyrics is rumored to improve a song's success on Christian radio; most hit CCM songs and on-stage banter feature terms such as "truth," "grace," and "salvation," as well as accounts of God as an active agent in one's life.

This evangelical theological outlook reflects the constituency of the so-called Christian market: the white evangelical Protestant Christians who comprise as much as 25 to 30 percent of Americans, or roughly 70 to 80 million people (Institute for the Study of American Evangelicals 2007). CCM tends to appeal to members of churches that originated in America (Baptist, Nazarene, Pentecostal, Free Methodist, nondenominational) rather than in Europe (Episcopal, Lutheran, Mennonite, United Methodist, Presbyterian, and Roman Catholic) (Powell 2002). Thus, the definition of CCM not only involves the characterization of Christian music, it also designates "Christian" to describe a particular kind of believer. This contributes to a peculiar American taxonomy in which Catholics, Eastern Orthodox Christians, Mormons and even mainline Methodists do not qualify as "Christian."

The lyrics and fans of CCM are particularly evangelical in part because CCM shares evangelical Christianity's focus on the conversion of the unsaved. The so called Great Commission (to spread the Christian faith) has fueled an engagement with new communications technologies, from Pat Roberston's Christian Broadcasting Network, to the film adaptations of the *Left Behind* book series, to CCM itself (Schultze 1990). Evangelical Christianity's separatist approach to culture also plays a role; since some evangelicals deem "mainstream" music unacceptable, music could only be performed and appreciated in an alternative form. While contemporary evangelical Christianity is particularly well-suited to produce and consume CCM, the most salient influence on the theology of CCM is its historical origin in the evangelical "Jesus Movement" of the late 1960s and early 1970s.

The CCM record industry

Most historians of CCM place its hearth in Southern California at the end of the 1960s. "Jesus Music" was the soundtrack of born-again hippies, known as the Jesus Movement, who rejected both the staid rigidity of mainline Christianity and the hedonism of the so-called youth counterculture (Thompson 2000). These "Jesus People" or "Jesus Freaks" infused the popular music of the day, slightly psychedelic rock, with their newfound evangelical Christian beliefs. The Jesus Movement found a home at churches such as Chuck Smith's nondenominational Calvary Chapel in Orange County. By the mid-1970s, Jesus Music began to cohere into a record industry, and by the mid-1980s the music was known as CCM—distinct from the longhaired hippie connotations of Jesus Music. Some Christians avoided the term "Christian rock" because rock was still considered to be demonic by a stalwart few.

In the early 1970s, Jesus Music appeared on gospel record labels (Word, Benson) and new Christian independent labels founded by churches (Calvary's Maranatha! Music). Major "secular" labels began to express interest in the nascent Christian market in the 1980s. A general-market music industry depression in 1979 (eventually ended by Michael Jackson's *Thriller* in 1983) fueled a need for new markets and new products. Major label interest in CCM was piqued by the platinum sales of Amy Grant's *Age to Age* in 1982, her crossover success in 1985, and her eventual sales of 22 million records (O'Donnell 1996; Rabey 2002). General-market success in the 1990s by CCM artists such as double-platinum alternative rock band Jars of Clay further enticed the major record companies. The revelation of CCM's expansive volume of sales by the introduction of the Christian Soundscan sales tracking system into bookstores in 1995 also attracted the majors (Stiles 2005).

The following flurry of buy-outs, mergers, and the creation of new labels left almost all of the CCM industry under the aegis of major general-market music corporations. Today, three of the "Big Four" major record companies control most CCM recordings. Warner Music Group, EMI, and Sony BMG own the top three CCM labels (Word, Sparrow, and Reunion, respectively) as well as many smaller CCM labels. The fourth major label, Universal, entered the fray by forming Universal Music Christian Group in December of 2006. EMI CMG Distribution (owned by EMI), Word Distribution (owned by Warner Music Group), and Provident-Integrity Distribution (owned by Sony BMG) distribute all significant CCM releases to stores. Even releases from "independent" Christian record labels, such as Tooth & Nail, are distributed to general market retail by the distribution wings of these three corporations.

Artists attempting to appeal to both the Christian and general markets will release the same album on two labels, sometimes called "joint distribution." One label markets to Christian retail and radio and another (often owned by the same parent corporation) targets the general market audience. For example, both general market label Capitol Records and CCM label Gotee Records released Relient K's *Five Score and Seven Years Ago* in 2007.

This corporate consolidation coincides with the more general music industry trend of major labels establishing internal divisions to appeal to niche markets (Bell 2003). It is also common for major labels to acquire a percentage of independent labels after the indie label has proven to be successful; even independent labels that do not sell a portion of the company to a major will often sign a distribution deal with a major distribution company. Economic geographer Allen Scott considers this system of joint ventures to be a way for major labels to cope with an unpredictable market (1999). Even though many people operating within the CCM industry value its potential for evangelism and worship, the involvement of the major music corporations in CCM has prioritized the search for new markets, return on investment, and an emphasis on short-term profits rather than long-term artist development (Romanowski 2000). However, as CCM scholars Jay Howard and John Streck argue, it is too simplistic to depict the corporate consolidation of CCM labels as co-optation (1999). Jerry Falwell's *Old Time Gospel Hour* and "chain" megachurches demonstrate that US evangelical Christianity has actively embraced media technologies, corporate organizational models, and for-profit business strategies. The logic of twenty-first century capitalism has not restricted the variety of musical styles available in CCM, nor has it diluted the religious content of CCM. It has, on the other hand, brought about an expanded geography of consumption and the centralization of production into an industrial district.

The improvement in recording quality, distribution, and marketing afforded by corporate consolidation has expanded the sales and popularity of CCM. Employing a definition of CCM that excludes traditional gospel, Southern gospel, and children's music, over 37 million CCM CDs, tapes, digital albums and digital tracks sold in 2006. This tally accounts for 69 percent of Christian/Gospel sales and 4.7 percent of all album sales, or $483 million (Gospel Music Association 2006; 2007).

The top sales markets for CCM are some of the country's largest metropolitan areas: Los Angeles, Dallas, New York, Chicago, and Atlanta. However, the top per-capita media markets for CCM are Omaha, Springfield (Missouri), Chattanooga, and Knoxville (Gospel Music Association 2006; US Bureau of the Census 2000). Indeed, the consumption of CCM assumes a regional pattern (Figure 16.1). Both total and per capita sales of CCM are strongest in the South and weakest in the Northeast. The South is also the part of the country where church congregations recently have grown the most, fueled by both Sunbelt migration and a propensity towards religious observance (Hadaway 2005). While CCM sales are proportionately higher in poor, white, rural, evangelical Christian, and Republican-voting media markets (Lindenbaum, forthcoming), the availability of CCM throughout the country suggests that corporatization has allowed CCM to become a nationwide phenomenon in both "red" and "blue" states.

As previously mentioned, most early CCM recordings were sold in Christian bookstores alongside bibles, fiction and nonfiction books, greeting cards, and assorted "Jesus junk" merchandise. The corporatization of CCM moved 64 percent of CCM sales to secular retail outlets, particularly "big box" stores such as Wal-Mart and Best Buy. CCM sales are also growing in Apple's online itunes store. As a result of this shift to big box and online retail, Christian bookstores now play less of a gatekeeping role in the CCM industry. Bookstore demand for sanitized, or safe, cover art and music is partly responsible for CCM's reputation

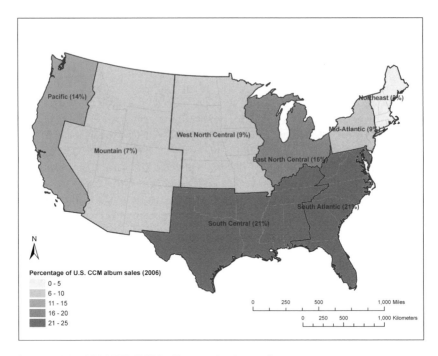

Figure 16.1   2006 US CCM album sales by region

of being aesthetically milquetoast (Beaujon 2006). Now, grandmother-unfriendly bands such as Underoath can sell their screamo via mp3s online or CDs in Virgin Records without fear of reprimand.

The corporate consolidation of the CCM industry has also contributed to the spatial concentration of production in Nashville, Tennessee. The first CCM industrial district was in Southern California, long a hotbed for novel forms of religious expression. At the same time that the CCM industry was being acquired by the major labels, artists and record companies moved to Nashville, Tennessee and its suburbs of Franklin, Brentwood and Hendersonville. Christian publisher Benson started the Greentree record label there in the 1970s; the Star Song record label moved to Nashville from Dallas in 1989, Sparrow from Southern California 1991, Word from Texas in 1992, and *CCM Magazine* from Southern California in 1989.

Many active CCM record labels are now concentrated in Nashville (Figure 16.2). Moreover, five of the six leading CCM record distributors in the United States are located in Nashville. The Nashville area boasts at least 42 CCM record labels, five CCM distributors, three CCM music publishers, two CCM radio networks, *CCM Magazine*, the Gospel Music Association, and Christian Copyright Licensing International.

Figure 16.2   CCM record labels in the US

While the geographic centrality of Nashville makes it appealing to touring acts, the primary attractions for the CCM migration to Nashville were the existing infrastructures of Protestant Christianity and the music and publishing industries. Sometimes known as the "Buckle of the Bible Belt" or the "Protestant Vatican," Nashville houses the headquarters of the Southern Baptist Convention, the United Methodist Church's Publishing House, religious publisher Thomas Nelson, and a large evangelical Christian community. Nashville supposedly has more churches per capita than any other US city (Daley 1998).

Nashville, renowned for its business-friendly climate, has long sustained industries that spin off related industries (Carey 2000). The Nashville area contains the studios, musicians, music publishers, producers, lawyers, and agents of the country and gospel music industries. Indeed, Nashville has the highest ratio of composers and musicians to total population in the country (Scott 1999); the Nashville area boasts over 20,000 songwriters (Daley 1998). EMI CMG head Bill Hearn credits the attraction of songwriters and music publishers for the industry's move to Nashville (Beaujon 2006). In Nashville, CCM artists and companies could produce better-sounding music for lower recording costs, while networking with the country music and publishing industries. Country stars Alan Jackson and Alabama found themselves on the Christian charts in 2006, and some Nashville

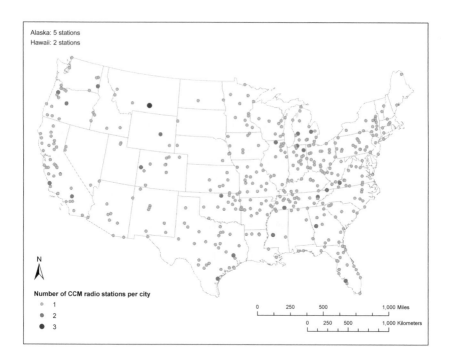

Figure 16.3   CCM radio stations in the US

record companies, such as Beach Street Records, operate in both the country and CCM industries.

The concentration of the CCM industry in Nashville demonstrates an agglomeration effect similar to the ones Allen Scott has noted in other parts of the music and film industries (Scott 1999; 2004). Agglomeration like this is reinforced by cultures, conventions, learning effects, and innovative impulses. The Nashville area offers a pool of human skills, infrastructure, as well as available capital; the CCM and country music industries consequently draw other artists, labels and industry professionals to Nashville.

CCM scholar William Romanowski (1990) argues that proximity to the Nashville music apparatus led to the formation of Amy Grant's "contemporary praise" CCM style in the 1980s. It is likely, however, that this spatial concentration in Nashville limits innovation as much as it fosters a vibrant creative field for both aesthetic and entrepreneurial creativity. The most stylistically and lyrically inventive forms of CCM tend to originate outside Nashville, while much Nashville-produced CCM ascribes to tried-and-true song structures, simple almost-clichéd lyrics, and radio-friendly studio polish. The spatial and industrial pressures on the music of CCM cannot be disassociated; both the corporate profit-orientation of the CCM industry and the concentrated spatial interaction of artists and executives reinforce these artistic tendencies.

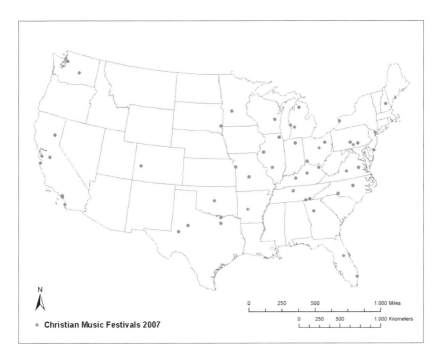

Christian Music Festivals 2007

Figure 16.4   CCM festivals in the US

CCM radio

Similar trends of an expanding geography of consumption and corporate conglomeration can be found in CCM radio. In 1974, there were no full-time CCM stations, and even as late as 1990, CCM radio networks and formats were rudimentary. As of 2007, the Gospel Music Association claims that over 20 million people listen to Christian/Gospel radio every week (Gospel Music Association 2007). While not all of these listeners tune in to CCM, the largest Christian format is Adult Contemporary (AC), and the sum of Christian AC, Christian Hits Radio, Christian Rock, and the Inspirational radio formats constitute the fourth largest format in the country. The continental United States contain at least 498 CCM radio stations, with an especially high concentration in the Upland South and the Midwest (Figure 16.3). Signal translators, satellite affiliates receiving syndicated content, and online radio allow CCM radio to reach many more places than those mapped. This rise mirrors that of Christian radio in general, which expanded its audience 43 percent from 2000 to 2005 (Piore 2005).

Religious, specifically Christian, stations are awarded most of the new noncommercial licenses by successfully tailoring their applications to the 1996 FCC guidelines. The largest owner of Christian radio is the non-profit Educational Media Foundation (EMF) with the fifth largest number of radio stations in the country overall (DiCola 2006). EMF owns 158 CCM stations; it claims to operate 303 signals for the Adult Contemporary format K-Love and 129 for the Christian Hits format Air-1 (some of which are affiliates or translator towers). EMF's K-Love claims to receive 96 percent of its money from listener support, and four percent from "other ministry support" (K-Love 2007). Another leading Christian radio company is the non-profit WAY-FM, which owns and operates 14 full power FM radio stations, 20 low-power repeater stations, and a satellite network. Some CCM stations are also for-profit; fourteen of these CCM stations are for-profit media conglomerate Salem Communications's "Fish" stations, targeting 25-54 year old Christian women in large radio markets.

CCM radio features the short playlists and clear musical identities typical of corporate-owned radio. Stations emphasize the "positive and encouraging" or "safe for the whole family" appeal of CCM, complete with extensive station branding. Even noncommercial stations dedicate airtime to self-advertisement and the promotion of their own, corporately-sponsored concerts. The homogeneity of CCM radio playlists, owners, and messages requires each individual station to produce a sense of local identity; the listener is addressed as a local listener and the centrally-owned stations are identified as belonging to a particular place (Savage 2002).

CCM festivals

Next to the recording and radio industries, festivals play a significant role in the CCM infrastructure. CCM concerts take place in arenas, churches, coffeehouses, and nightclubs, but huge multi-day summer music festivals are the most important venues for live performance. Many festivals are located close to amusement parks, feature nonmusical Christian speakers in addition to musical artists, and offer discounted pricing for youth groups.

In 2007, at least 54 CCM festivals took place in the United States (Figure 16.4), many of them over extended summer weekends. Tens of thousands of teens, parents, and youth group leaders attended CCM festivals in 2007. Celebrate Freedom in Parker, Texas, claims to draw 210,000 people to the largest free outdoor concert in the country, and the Creation festival spawns a tent city of 80,000 in south-central Pennsylvania (Sheler 2006). These festivals cater almost exclusively to CCM fans who already consider themselves to be Christians (Howard and Streck 1999).

The festivals are not necessarily located in or near hubs of CCM album sales. There is no significant statistical relationship between a media market having a festival and its 2006 CCM per capita sales (Lindenbaum 2009). Indeed, seven of the top 20 CCM sales markets do not have a CCM festival. Rather than aiming for a strong Christian music fan base, festival organizers base their event in their home town, in an area that lacks a Christian music festival, or in the place in which "God called them" to host the festival.

The pre-existing venues that house these festivals (i.e. farms, racetracks, amusement parks, vacation recreation areas) require large expanses of land and are rarely located in dense urban areas. Many Christian music festivals include camping for families and youth groups, which requires a recreation area. Large parking lots or fields for parked cars also necessitate large open spaces. Since many festivals draw fans who travel with automobile from bordering states and even other countries, they do not need to be in a city center with access to public transportation.

The cultural progeny of revival camp meetings, these music festivals feature explicitly Christian discussions, T-shirts sales, booths for radio stations, and tables for organizations such as anti-abortion Rock for Life and child sponsoring Compassion International. The sound of the music transforms the landscape of a raceway or theme park—spaces that would rarely be considered to be Christian—into an area of community and faith, where raising one's hands up in praise is considered the norm rather than eccentric. Banners hang, T-shirts proclaim, stages resound, audiences pray, and the theme park, raceway, fairgrounds, or farm becomes a specialized arena for the expression and reinforcement of evangelical Protestant Christianity.

The secular/sacred dualism

Geographer Allan Pred writes, "to categorize is to define. It is to demarcate, to indicate not only what something – or set of things – is, but also what it is not … it is – to (attempt to) exercise power" (1995, 1067). The delimiting inherent in the creation, appreciation, and even academic analysis of CCM produces musical meanings, Christian consumers, and notions of what it means to be Christian. As CCM artist and producer Charlie Peacock writes, "All ideas have consequences, including the idea of naming things Christian" (Peacock 1999, 136).

Since its inception, CCM has been depicted as a parallel industry or entertainment universe (Ali 2001), an element of a rural and Republican semi-autonomous evangelical subculture that appeals only to the already-converted through a particular evangelical vocabulary. The rhetoric of a divide between secular society, or "the world," and a sacred Christian subculture can be traced back to the 1925 Scopes trial and the system of Christian schools, colleges, churches, camps, publishers, and organizations that developed afterwards (Balmer 1989). The very utterance of "CCM" posits the existence of a bounded and internally coherent set of Christian voters or consumers. The particular geography of CCM—its sales patterns, industrial concentration in Nashville, CCM-only music festivals, and CCM-only radio stations—reinforces this dichotomy. However, at closer inspection, CCM is not organizationally or geographically distinct from a separate secular or mainstream "world." The CCM industry is not parallel to and separate from the so called mainstream record industry; many of the parent corporations, business strategies, and retail locations are the same. Many Christian musicians with general market appeal choose to disassociate themselves from the CCM industry, and most CCM fans also listen to non-CCM artists. Even within the CCM industry, there has been some resistance to this secular/sacred dualism. In the May 2007 issue of *CCM Magazine*, editor Jay Swartzendruber announced, "we're going to stop perpetuating the myth that what is and is not 'Christian music' is based on where the music is sold" (2007, 4). CCM journalist Mark Joseph is actively critical of the self-imposed separatism that led to CCM receiving the label of "musical ghetto" (Joseph 1999). We are at a transitional moment in the history of CCM—corporate consolidation and the blending of what once may have been considered separate "secular" or "Christian" cultural practices have arguably rendered secular/sacred distinctions untenable.

The secular/sacred divide is a socially produced marketing distinction, but not one without social impact. As I will explore in future research, CCM has the potential for profound social effects—the reinforcement of faith, the formation of community sentiment, the mobilization of non-profit organizations and political parties, and the production of Christian political worldviews. Rather than adopt a rhetoric of cultural and political polarization, we should understand the category of CCM and that evangelical America is in a constant state of change.

References

Ali, L. (2001), "The Glorious Rise of Christian Pop," *Newsweek* 16 July, p. 38.

Balmer, R. (1989), *Mine Eyes Have Seen the Glory: A Journey into the Evangelical Subculture in America* (New York: Oxford University Press).

Beaujon, A. (2006), *Body Piercing Saved My Life: Inside the Phenomenon of Christian Rock* (Cambridge, MA: De Capo Press).

Bell, T. (2003), "Why Seattle? The Origins of an Alternative Rock Culture Hearth," in George O. Carney (ed.).

Carey, B. (2000), *Fortunes, Fiddles & Fried Chicken: A Nashville Business History* (Franklin, TN: Hillsboro Press).

Carney, G.O. (ed.) (2003), *The Sounds of People and Places: A Geography of American Music from Country to Classical and Blues to Bop* (Lanham, MD: Rowman and Littlefield).

Daley, D. (1998). *Nashville's Unwritten Rules: Inside the Business of Country Music* (Woodstock, NY: Overlook Press).

DiCola, P. (2006), *False Premises, False Promises: A Quantitative History of Ownership Consolidation in the Radio Industry* (Washington, DC: Future of Music Coalition).

—— (2007), U.S. Radio Station Data (Washington, DC: Future of Music Coalition).

Forbes, B. and Mahan, J. (eds.) (2000), *Religion and Popular Culture in America* (Berkeley: University of California Press).

Gospel Music Association (2006), National DMA/Genre Sales Data, (Nashville, TN).

Gospel Music Association (2007), "Christian/Gospel: Music that Connects – Industry Overview 2007," (Nashville, TN).

Hadaway, C.K. (2005), *Facts on Growth* (Hartford, CT: Faith Communities Today and the Cooperative Congregational Studies Partnership).

Hiatt, B. and Serpick, E. (2007), "The Record Industry's Slow Fade," *Rolling Stone* 28 June, p 13-14.

Howard, J.R. and J.M. Streck (1999), *Apostles of Rock: The Splintered World of Contemporary Christian Music* (Lexington, KY: The University Press of Kentucky).

Institute for the Study of American Evangelicals (2007), "How Many Evangelicals Are There?" <http://www.wheaton.edu/isae/defining_evangelicalism.html>, accessed 14 May, 2007.

Joseph, M. (1999), *The Rock & Roll Rebellion: Why People of Faith Abandoned Rock Music – and Why They're Coming Back* (Nashville, TN: Broadman & Holman Publishers).

K-Love (2007), "Listener Information Pack," <http://www.klove.com>, accessed 21 June, 2007.

Lindenbaum, J. (2009), *The Industry, Geography, and Social Effects of Contemporary Christian Music* (Berkeley: University of California, Ph.D. dissertation).

O'Donnell, P. (1996), "Rock of Ages," *The New Republic*, 18 November, p. 14.

Peacock, C. (1999), *At the Crossroads: An Insider's Look at the Past, Present and Future of Contemporary Christian Music* (Nashville, TN: Broadman & Holman Publishers).

Piore, A. (2005), "A Higher Frequency: How the Rise of Salem Communications' Radio Empire Reveals the Evangelical Master Plan," *Mother Jones* December, 47-51, 80-1.

Powell, M.A. (2002), *Encyclopedia of Contemporary Christian Music* (Peabody, MA: Hendrickson).

Pred, A. (1995), "Out of Bounds and Undisciplined: Social Inquiry and the Current Moment of Danger," *Social Research* 62:4, 1065-1091.

Rabey, S. (2002), "A Chastened Singer Returns to Christian Basics," *New York Times*, 11 May, p. 14.

Radosh, D. (2005), "The Devil's Music," *salon.com*, 24 November.

Romanowski, W. (1990), *Rock 'n' Religion: A Sociocultural Analysis of the Contemporary Christian Music Industry* (Bowling Green, OH: Bowling Green State University, Ph.D. dissertation).

—— (2000), "Evangelicals and Popular Music: The Contemporary Christian Music Industry," in B. Forbes and J. Mahan (eds.).

Savage, T. (2002), "Christian Radio and the Metropolitan Edge" (Berkeley: University of California).

Schultze, Q. (ed.) (1990), *American Evangelicals and the Mass Media: Perspectives on the Relationship between American Evangelicals and the Mass Media* (Grand Rapids, MI: Academie Books).

Scott, A. (1999), "The US Recorded Music Industry: On the Relations Between Organization, Location and Creativity in the Cultural Economy," *Environment and Planning A* 31, 1965-84.

—— (2004), *On Hollywood: The Place, the Industry* (Princeton, NJ: Princeton University Press).

Sheler, J.L. (2006), *Believers: A Journey into Evangelical America* (New York: Viking).

Stiles, J. (2005), "Contemporary Christian Music: Public Relations amid Scandal," *Journal of Religion and Popular Culture* 11: 3.

Styll, J. (2007), E-mail communication, 22 March.

Swartzendruber, J. (2007), "Seeing Is Believing," *CCM Magazine* May, p. 4.

Thompson, J.J. (2000), *Raised by Wolves: The Story of Christian Rock & Roll* (Toronto: ECW Press).

US Bureau of the Census (2000), <http://factfinder.census.gov>, accessed 15 April, 2008.

Index

The compilation of the index was assisted
by Lori Kimmick at the University of
Pittsburgh at Johnstown.